W9-BMI-396

Michael W. Dempsey Editor-in-Chief
Angela Sheehan Executive Editor
Peter Sackett Production Manager
Marion Gain Picture Researcher
Sarah Tyzack Picture Editor

ISBN 0 361 05319 3

Published 1983 by Purnell Books, Paulton, Bristol BS18 5LQ
a member of the BPCC group of companies.
Made and printed in Great Britain by Purnell and Sons
(Book Production) Limited, Paulton, Bristol

Purnell's Pocket Concise Encyclopedia of Nature

Adapted from Purnell's Concise Encyclopedia of Nature
by Robert W. Burton

Original text by Michael Chinery

Purnell

Contents

Living Things

Nature or natural history is sometimes taken to mean the study of natural things, or, more usually, the study of living things. Scientists often use the word *organism* for anything that is alive. Trees and flowers, butterflies and tortoises, fishes and elephants are all living things or organisms. Rocks, metals, and water are not organisms.

The first thing we must do if we are going to study living things is to find out what it is that makes living things different from non-living objects. We do not know the vital ingredient necessary to make something alive, but we can list seven features possessed by living things.

Characteristics of Life

One of the most obvious features of living things is that they can *move* by themselves, in search of food perhaps or to escape from their enemies. *Feeding* (including drinking) is another feature of all living things. They have to take in food to provide themselves with energy and with body-building materials. Energy is released from food when food is 'burnt' or combined with oxygen, and so living things have to take in oxygen from the air. The process of taking in oxygen is called *breathing* or respiration. Body-building materials in the food are taken to all parts of the organism's body and added to it, so that it *grows*. Breathing and the other processes that go on inside the living organism result in the production of waste materials. These must be removed, and so all organisms have some way of *excreting*, or getting rid of these waste materials. Living things are also *sensitive* to changes in their surroundings. A fly will soon be off if it senses the approaching fly swatter, and a potted plant will soon turn upwards again if its pot is laid on its side. Finally, all living things can *reproduce* themselves. Organisms cannot live for ever because their bodies gradually wear out, but they ensure that more of the same kind will take their place by producing seeds or eggs or babies. A few simple organisms reproduce by splitting in two.

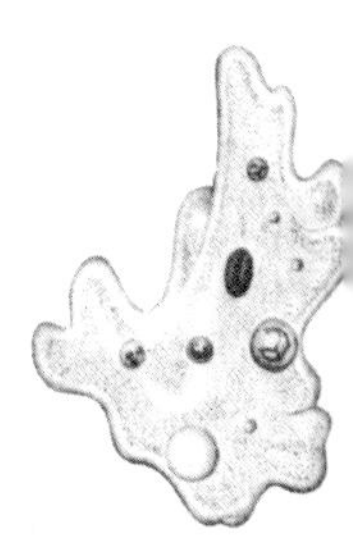

Amoeba, a tiny speck of jelly, is an animal with all the characteristics of living things.

Even when taking a rest this serval cat is alert and sensitive to what is going on—an important feature of living things.

Plants are just as much alive as animals. You can watch them growing and turning towards the light.

The Living Cell

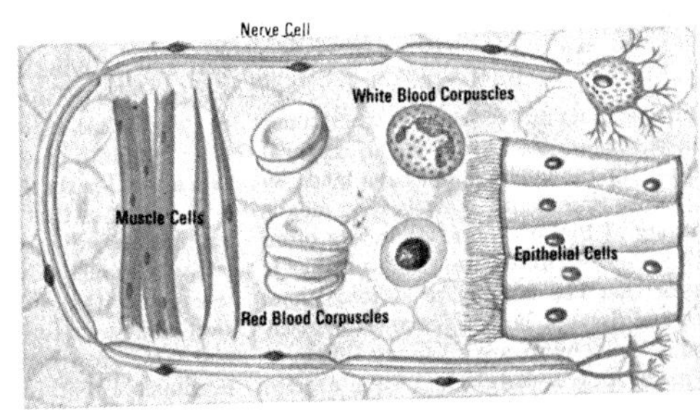

Just as a house is built of bricks, so animals and plants are made of units called cells. Cells are extremely small and they cannot be seen without a microscope. Some animals and plants —known as the Protozoa or Protista— consist of only one cell, but larger creatures consist of many, many cells. There are different kinds of cells and each kind has a particular job to do. There are skin cells; nerve cells carrying messages from place to place; bone cells forming our skeletons; and muscle cells enabling us to move. Plants have a variety of cells too. There are food-making cells in the leaves, tubular water-carrying cells, and food-storing cells.

Animals and plants are built of cells. The cells are not all the same shape. Different shaped cells have different jobs to do. The diagram above shows a selection of human body cells.

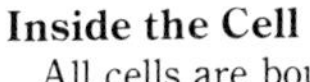

Plants and Animals

Plant cells differ from animal cells mainly in having a stiff cellulose wall around them. This gives them a fixed, regular shape. Animal cells do not have cell walls and they are less rigid than plant cells. Plant cells also normally contain a large sap-filled space called a vacuole. Plant cells often contain small green objects called chloroplasts. These give them their green colour and they also help the plants to make food.

Although cells are so tiny, they have complicated structures inside them. Fine channels carry chemical materials from the nucleus to all parts of the cell.

Inside the Cell

All cells are bounded by a very thin cell membrane. In plants it is just inside the cell wall. Inside the membrane is the protoplasm and in the centre of most living cells there is a darker, dense region called the nucleus. This is made of a special type of protoplasm and is the 'brain' of the cell. Chemical materials made in the nucleus spread out into the protoplasm through minute channels and they 'tell' the protoplasm what to do.

Inside the nucleus are the chromosomes — minute thread-like structures which carry the genes. Genes are very complicated chemical materials and each one carries the 'plans' for a part of the body. The nucleus instructs its cell to make things according to these 'plans', and so the whole body is made up in the correct way.

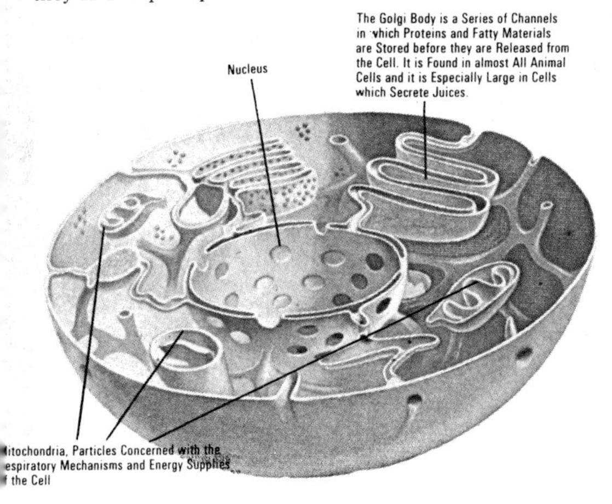

Cell Division

Living things normally start out as a single cell, called an egg cell, which divides into two. The two cells then divide into four, and so it goes on. Cell division of this sort takes place even in the fully grown plant or animal because new cells have to be produced to replace those that have worn out. The chromosomes play a very important part in cell division because it is essential that the new cells have the same 'instructions' as the old ones. Each chromosome produces an exact copy of itself and then, when the cell divides into two, one set of chromosomes goes to each new cell.

The Origins of Life

When the Earth first came into existence some 4,500 million years ago there was no life on the planet. Starting probably as a dense ball of dust, the Earth passed through a very hot and partly liquid stage. Volcanoes belched out carbon dioxide and the atmosphere also contained large amounts of methane (marsh gas), ammonia, and steam. Things gradually cooled down and much of the steam turned into water and formed the ocean.

Virus particles come in many different shapes, as shown by the two types of 'flu' virus (top) and the smallpox virus.

Viruses

Viruses are minute germs not much more than one millionth of a millimetre across. They appear to be on the borderline between living and non-living matter. Viruses can grow and reproduce only inside living cells. The virus 'instructs' the cell to make more virus material instead of normal cell material. This interferes with the normal working of the cell and produces disease in the animal or plant. The damaged cells eventually release the viruses which move on to other cells in the body. Foot and mouth disease, fowl pest, and myxomatosis are among the many animal diseases caused by viruses. Plant diseases caused by viruses include tobacco mosaic, and many others.

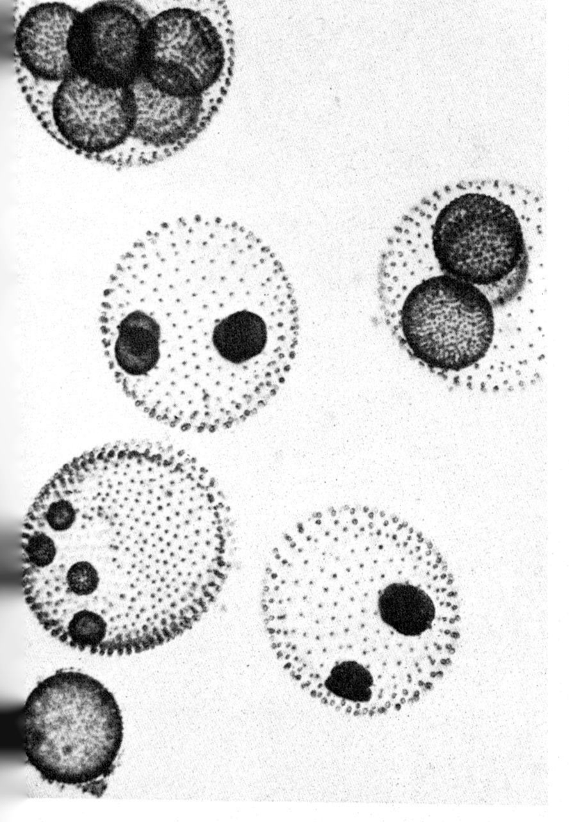

The Primeval Soup

How did life start on the Earth? We shall never know exactly how it happened. The atmosphere of our planet in its younger days consisted largely of methane, carbon dioxide, ammonia, and water. These are simple substances but they contain all the major elements of life—carbon, oxygen, hydrogen, and nitrogen. The sun's ultra-violet light was able to get through this atmosphere and, under the influence of this radiation, some of the simple materials probably linked up to form sugars and amino-acids. These substances are the building bricks of living material.

We can only guess at the next stage in the development of life, but presumably some of the sugars and amino-acids linked up and produced proteins and other complicated materials that could actually duplicate themselves. If these materials then formed some sort of membrane around themselves they would have produced the first cell. It would have taken a long time for the right combination of materials to appear, but it could have happened. In fact, recent reports suggest that scientists have succeeded in producing a living cell.

Microscopic algae may have been an early form of life.

Plants and Animals

We do not know what the earliest living things were like, and we probably never will. They might have been something like today's bacteria. Whatever they were like, we can be fairly certain that they have given rise to all of today's living things with the possible exception of the viruses.

Apart from the viruses, all of today's living things fall into one or two large groups—plants and animals. Examples of plants include mosses, seaweeds, trees, grasses, and cabbages. Examples of animals include jelly-fishes, spiders, snakes, birds, and dogs. Man is also an animal.

It is not so easy to tell the difference between all plants and animals. You might think that animals move about and plants stay still, but this is not always so. There are many animals, such as the sea anemone, that stay in one place and there are many tiny plants that swim about in the water. Then there is the insect-eating sundew—a plant whose sticky 'fingers' can move quickly to trap a small fly.

The most important distinction between the two groups concerns the way in which they get their food. Nearly all plants are able to make their own food, and those that cannot do this are obviously plants from their other features. Animals cannot make food at all and they always have to feed on plants or upon other animals that have themselves eaten plants.

It is easy to distinguish a typical plant, such as a buttercup, from a typical animal . . .

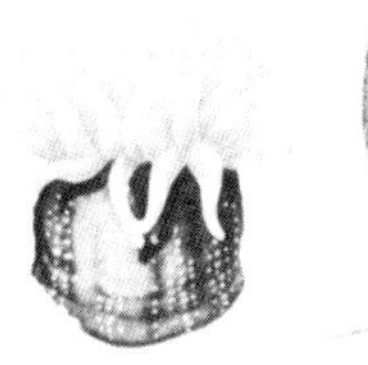

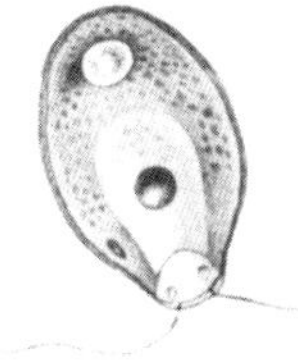

. . . but less easy to say which of these is a plant and which is an animal. The sea anemone (left) is an animal, and the other is a microscopic plant.

Plant and Animal Species

Each of the many different kinds of animals and plants is known as a species. There are probably well over a million different kinds of animal species, and something like half a million plant species.

We can define a species, therefore, as a group of plants or animals which all have the same appearance and behaviour and which can breed among themselves.

We know today that the appearance and behaviour of an organism depend upon and are controlled by the genes present in its cells. All members of the species have the same arrangement of genes and chromosomes in their cells. When the animals mate or the plants are fertilized this same pattern is passed on to the next generation, and so they will also conform to the general appearance and behaviour of the species. The structure and behaviour of different animals normally prevents them from mating other than with members of the same species.

Closely related species have many features in common because they have evolved fairly recently from a common ancestor. The picture shows a great tit (left), a varied tit (top right), and a blue tit (bottom right).

Species and Genera

Linnaeus, the distinguished Swedish naturalist of the eighteenth century, was one of the first people to try to classify things. He recognized that some species were very alike and grouped them into larger units called genera (singular genus). He then gave a scientific name to each species which was made up of the name of the genus and the name of the species. The scientific name of the large white butterfly is *Pieris brassicae,* while that of the small white butterfly is *Pieris rapae.*

The Species and Evolution

Although all the members of a species look basically alike, there are always some minor variations. You have only to look at people to see this variation: they all belong to the same species, but none of them look *exactly* alike. The same thing occurs among other animals and plants. When all the members of the species are free to breed among themselves these small differences are evened out and the species stays more or less the same. But suppose that the members of a species become separated into two regions—by a chain of mountains for example. Very small changes may occur in one population and not in the other. Over a very long time these changes might spread throughout one population and it would be slightly different from the other population. They would still belong to the same species, and would probably be known as geographical races or varieties of that species. If the two populations continued to be separated, the differences between them might well increase until there came a time when they were no longer able to interbreed. The two populations would then be separate species. This sort of process has given rise to all the many species we know today. Species which are very similar to each other are those that have been separated most recently.

Although one species does not normally mate with another, closely related species do sometimes mate and produce offspring. These offspring are called *hybrids*. One of the best known examples is the mule, the offspring of a male donkey and a female horse. The mule is sterile and unable to reproduce.

Plant Classification

Unlike most animals, plants do not have a definite shape and no two plants will ever look exactly alike. But the members of one species are all quite similar. Two bluebell plants have more in common than a bluebell and a daffodil. There is a closer connection between a bluebell and a daffodil, however, than between a bluebell and a primrose. And there is a closer connection between a bluebell and a primrose than between a bluebell and a fern. These similarities and differences are used to classify plants, or to divide them into groups.

Most people who study plants divide the plant kingdom into four major groups. They are called thallophytes, bryophytes, pteridophytes, and spermatophytes.

Thallophytes

The thallophytes are the simplest of the plants alive today, and the earliest plants must have been of this type. There are no flowers, and the plant body is not even divided into root, stem, and leaf. There are two main groups: the Algae and the Fungi. Bacteria are usually classified as thallophytes as well, although they are rather different from the other members of the group.

Bryophytes

The bryophytes are green plants which usually have distinct stems and leaves. There are no true roots, however, and no flowers. Bryophytes are divided into two main sections: the mosses and the liverworts.

Pteridophytes

This group includes the ferns, horsetails and club mosses. They are green plants and they have stems and leaves as a rule, and they also have proper roots. Associated with their larger size, they also have special tubes which carry water and food around the plant. There are no flowers.

The leaves of a monocotyledon such as an iris are normally narrow with parallel veins. The leaves of a dicotyledon such as a rose are normally broader and have a network of veins.

Spermatophytes

These are the seed-bearing plants. Roots, stems and leaves are normally well developed. Apart from a few parasites (see page 35), they contain chlorophyll and make their own food. There is a tremendous variation in size in this group, from minute floating duckweeds to huge redwood trees.

The two major sections of the seed-bearing plants are the gymnosperms and the angiosperms. Most of the gymnosperms carry their seeds in cones and are therefore called conifers. Other gymnosperms include the cycads, the yews, and the ginkgo or maidenhair tree.

The angiosperms are the flowering plants. They account for something like 250,000 of the known plant species. There are two main sections: the monocotyledons and the dicotyledons. The basic difference between the two sections is found in the seed. The seed of a monocotyledon has only one seed leaf (cotyledon), whereas the seed of a dicotyledon has two seed leaves. Monocotyledons include plants such as grasses, irises, daffodils, tulips, and orchids. Their leaves are normally narrow, and the veins run parallel to each other. The flowers frequently have either three or six petals. The dicotyledons normally have broader leaves, with a network of veins. Their flowers usually have either four or five petals. Almost all flowering trees are dicotyledons, and so are most garden plants.

Within the monocotyledons and dicotyledons, the plants are arranged in many families. The arrangement depends very much on the structure of the flower. The leaves and stems are not of much use in classifying plants because they vary so much according to the conditions under which the plants live.

CYCADS
CONIFERS
GINKGOS
FERNS
CLUBMOSSES
HORSETAILS
MOSSES
LIVERWORTS
FUNGI
BACTERIA
ALGAE
GYMNOSPERMS
ANGIOSPERMS
DICOTYLEDONS
THALLOPHYTA
BRYOPHYTA
PTERIDOPHYTA

Bacteria

Bacteria are minute organisms, usually regarded as very simple plants, although they do not contain chlorophyll. The largest bacteria are only about 0.01 millimetre across. There are three main groups of bacteria, known as bacilli, cocci, and spirilli. Bacilli are rod-shaped, cocci are spherical, and spirilli are like minute curved or twisted sausages.

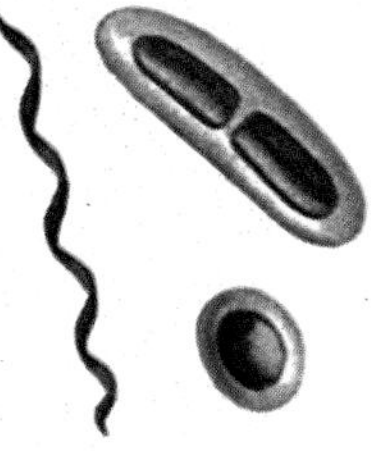

We often think of bacteria as disease-causing organisms or germs. Many certainly do cause diseases, such as tuberculosis and tetanus, but the majority are very useful. Bacteria in the soil get their energy by breaking down dead plants and animals. In doing so, they release minerals which growing plants can take up again. Bacteria therefore play a vital role in the economy of nature.

The Algae

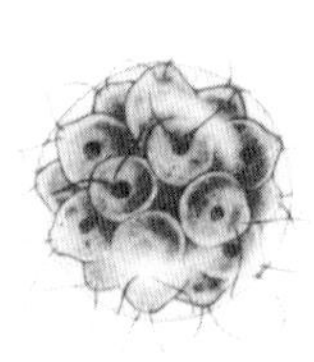

Above: *Pandorina*, a colonial alga consisting of 16 individual plants embedded in a sphere of jelly.

The algae are simple plants belonging to the large group known as the thallophytes. The plant body is called a *thallus*. It is not divided into roots, stems, and leaves, although some of the seaweeds have root-like holdfasts which fix them to the rocks. All the algae possess chlorophyll, although it is often masked by brown or red pigments.

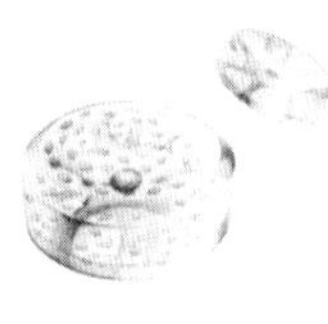

A diatom, one of the algae with glassy shells.

Plants That Move About

The simplest algae are single-celled organisms and most of them are extremely small. Some of them live in damp soil and some of them clothe tree trunks in damp places, but most of them live in water. They possess little whip-like hairs called flagella, and they swim by lashing them.

One of the commonest algae in fresh water is called *Chlamydomonas*. Its single oval cell is so tiny that fifty of them would only just about stretch across a pin's head. But this minute creature is a complete plant. It has a cellulose cell wall, a nucleus to control its activity, and a chloroplast in which it can make food. The little plant also has a red 'eye spot' which is sensitive to light. This eye spot controls the movement of the plant, directing it towards areas of moderately bright light. This ensures that the plant remains in a position where it can carry out photosynthesis efficiently. When the water is warm *Chlamydomonas* grows rapidly and it splits into two every day or so.

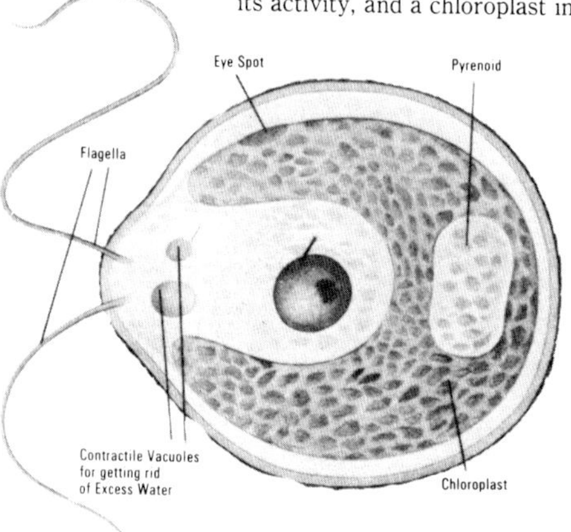

Chlamydomonas, greatly magnified, showing the cup-shaped chloroplast and the two flagella. The dark sphere in the centre is the nucleus.

There are several common pond-dwelling algae which live in little clusters or colonies. The colonies consist of eight or more cells, all more or less like *Chlamydomonas.* Although the cells are all loosely connected in a mass of jelly, each acts as a separate individual.

Above: *Spirogyra*, showing the spiral chloroplast which gives the plant its name.

Spirogyra

During the summer many ponds become choked with tangled green strands known as blanketweed. Each strand is an alga, and one of the commonest kinds is called *Spirogyra.* The strands are unbranched and each is composed of a number of cylindrical cells. These cells are all more or less alike and each acts independently, making food in its spiral chloroplast. It is this spiral chloroplast that gives the plant its name. The cells divide into two every now and then, but the two new cells do not separate: they remain in the strand and continue to grow. The strand therefore increases in length.

Above: The algae are prominent on rocky sea shores as seaweeds. They provide a retreat for many small animals when the tide goes out.

Below: A few of the many varieties of seaweed.

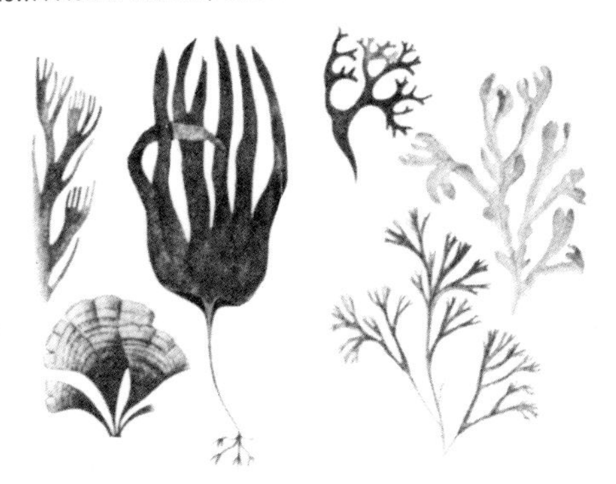

Seaweeds

Seaweeds include the largest and the most complicated algae. They are red, brown or green. The need for sunlight means that the seaweeds can grow only in shallow water, where light can reach the bottom.

Green seaweeds, such as the common sea lettuce, normally grow on the upper parts of the sea shore and they make the rocks very slippery. A little lower down the shore there is usually a zone of brown seaweeds. These include the largest of the algae, some of the oarweeds or kelps being many metres long. One of the commonest of the brown seaweeds is the bladder wrack, so called because its fronds bear numerous air bladders which pop if squeezed. The fronds of the brown seaweeds are covered with pale spots, especially at the tips. These are tiny pits in which the reproductive bodies develop. Tiny free-swimming spores are released from the pits. Some of them have to pair up before growing into new seaweeds, but others can grow directly into new plants.

Red seaweeds are generally much smaller than the brown ones. Instead of having broad blades, many of them have slender branching filaments. Some secrete chalky skeletons around themselves and these seaweeds play an important part in building coral reefs. Red seaweeds normally grow further down the shore than the brown ones and, unless there are rock pools, they do not very often reach above low tide level. They cannot withstand exposure to the air as well as the other seaweeds.

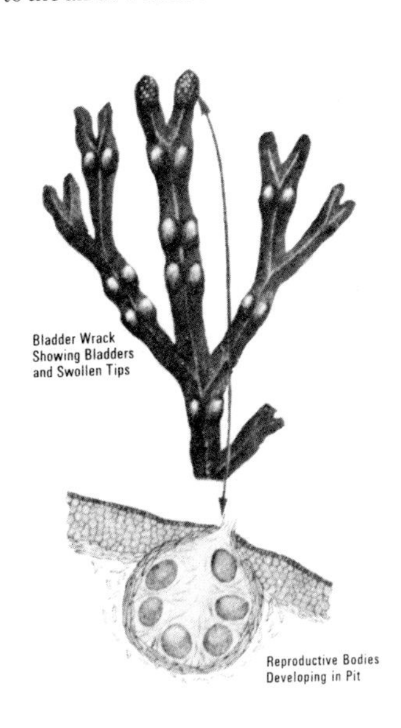

The bladder wrack is one of the commonest brown seaweeds. The bladders contain air which helps to buoy up the plant. The tips of the thallus bear pimply swellings. Each pimple is the opening of a reproductive organ.

Fungi

Moulds and toadstools belong to the large group of plants called Fungi. Some of them are clearly similar to the algae, but none of the fungi ever has any chlorophyll. Fungi cannot, therefore, make their own food. Most of them get it from dead and decaying matter, such as dead leaves and rotting wood. Fungi getting food in this way are called *saprophytes*. Some fungi, however, are *parasites* and they obtain food from living plants or animals.

Because the fungi do not make their own food they do not need sunlight. This is why we can grow mushrooms in dark damp cellars. Like the algae, the fungi belong to the division of plants called the thallophytes. They have no roots, stems, or leaves. Their bodies are composed basically of slender threads called hyphae. These creep over or through the food materials and secrete digestive juices which dissolve the food. Fungi are of great importance in nature. They play a major part in breaking down dead leaves and returning food materials to the soil.

Moulds

Moulds are relatively simple fungi. They consist of a mass of fluffy threads that cover the material they are growing on. One of the common moulds is the pin mould, scientifically known as

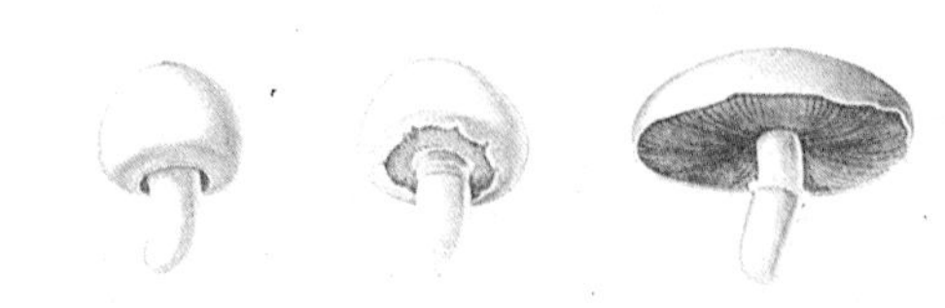

Above: The mushroom begins as a small, round cap and grows into the familiar parasol shape.

Above: The beautiful but poisonous fly agaric.

Few organic materials are safe from the attacks of fungi. Mucor, the pin mould, grows on stale bread, cheese, leather, animal dung, and many other substances. The pictures (below, left) show the little black capsules which contain the spores greatly enlarged.

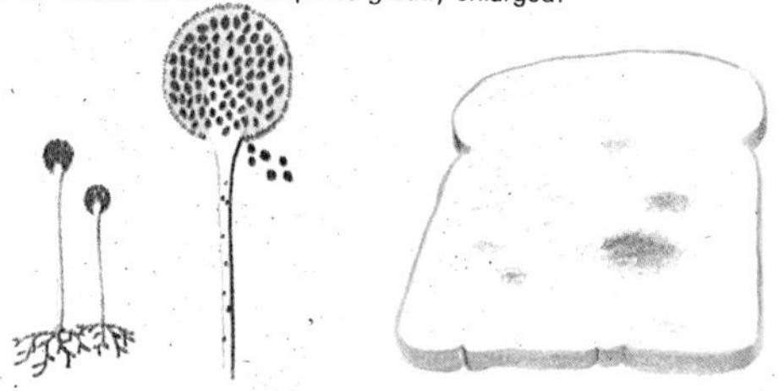

Penicillin

Penicillin was discovered quite by accident in 1928, when Alexander Fleming found that a little green mould was growing among some bacteria (germs) he was studying at a London hospital. Fleming then noticed that the germs near the mould were dying. In fact, the mould was producing a germ-killing substance.

Yeasts

Yeasts are rather special kinds of fungi which consist of tiny egg-shaped cells instead of threads. There are many different kinds and they can turn sugar into alcohol. Yeasts are therefore used in making wine and beer.

Mucor. It grows on a variety of materials and can be grown very easily on a piece of damp bread on a saucer. Like nearly all fungus threads, the hyphae are white, but upright branches form after a while and they bear little black capsules at their tips. These look like tiny pinheads and give the mould its common name. Each capsule contains hundreds of tiny spores and, when the capsule splits, the spores are scattered like dust into the wind. If they land in a suitable place they will grow into new threads. This is a form of vegetative or asexual reproduction. Sexual reproduction takes place when side branches join together and form tough-walled 'eggs'. These can survive cold and drought and, when good conditions return, they start to grow. A spore capsule develops and scatters lots of spores which can then grow into new threads.

Toadstools

Toadstools are more complicated fungi than the moulds, but the familiar umbrella-shaped toadstool is only a part of the fungus. For most of the year the fungi exist as slender threads just like those of the moulds. They are usually under the ground and many of them are associated with tree roots. They grow and branch. Then they become tightly packed together and they start to form the toadstool. At first this is a little button-shaped object in the ground. It grows rapidly and breaks through the surface. The cap then opens out ready to release the spores.

The mushroom is simply one particular kind of toadstool that happens to be very good to eat. There are many other toadstools that are good to eat, but some of them are very poisonous. You should never eat fungi unless you are absolutely sure that they are edible. Two of the most deadly kinds are the death cap and the red and white fly agaric. Both have little 'cups' at the base of the stalk.

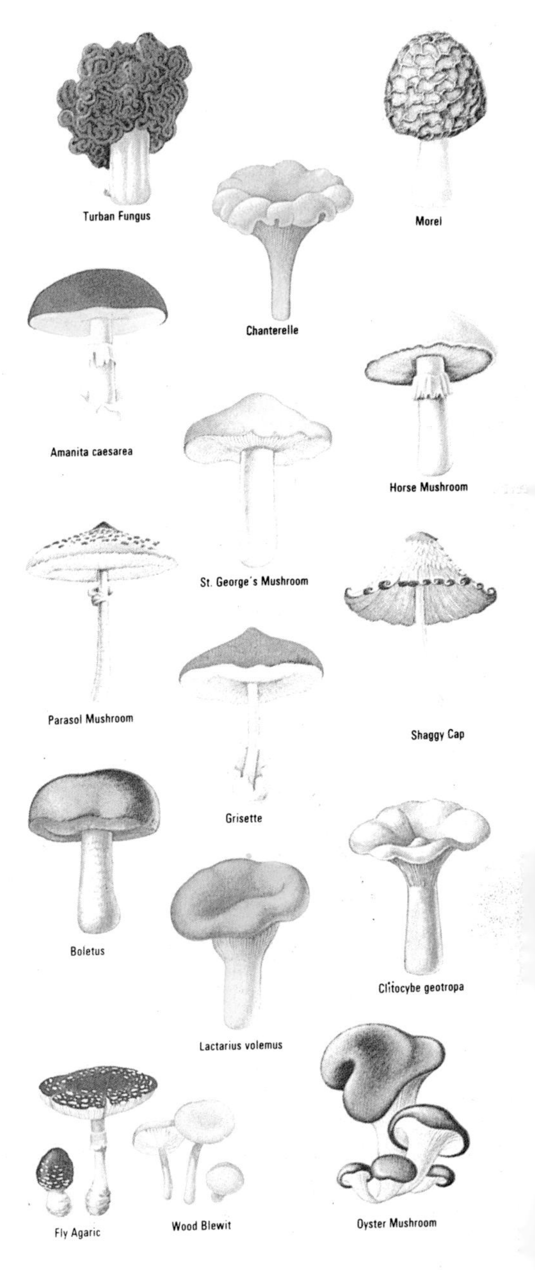

Lichens

Old walls, especially in country areas, are often encrusted with grey or yellow patches more or less circular in outline. These patches, despite their rather lifeless appearance, are growing plants and they are called lichens. They are extremely hardy plants and their hardiness results mainly from their curious construction. A lichen is, in fact, two plants rolled into one. The two plants are a fungus and an alga, each kind of lichen having its own combination of fungus and alga species.

The bulk of the lichen plant is made up of fungus threads, but there is a great difference between the tough, hardy lichen and the soft body of a normal fungus. This is one of the many mysteries still surrounding the lichens. The tiny algae are bound up among the fungus threads and are normally concentrated near the upper surface where they can receive the maximum amount of sunlight.

The fungus threads of the lichen absorb and store water, then pass it on to the algae. The algae contain chlorophyll and carry out photosynthesis. Mineral salts are obtained partly from the dust that falls on the plants and partly from the rock or soil. The minerals and sugars are combined by the algae to make food for themselves and for the fungus. The lichen is therefore very much a partnership. It is a good example of *symbiosis* (see page 85).

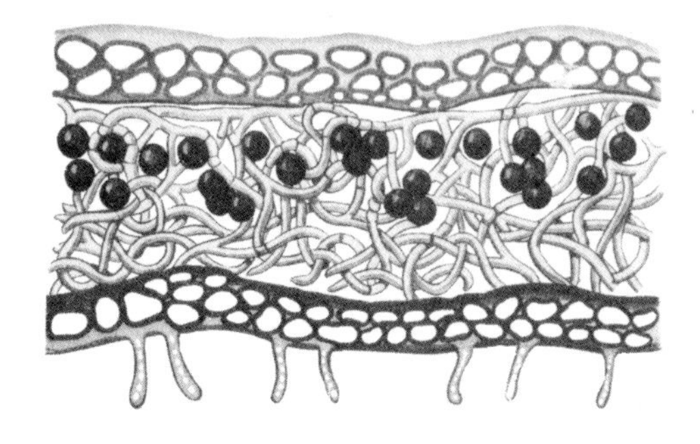

The bulk of the lichen body is formed by the tangled fungus threads. The algal cells are found mainly in the upper layers, where they get enough light to make food.

Reproduction takes place in two ways in the lichens. The fungus produces tiny spores which are scattered in the wind but they contain no algae and will not develop into lichens unless they meet the right kind of alga. The other method of reproduction involves the production of *sporedia*. These are little granules that break off the lichen surface. They contain fungus threads and algae and they grow directly into new lichens.

There are three main groups of lichens. Right: *Foliose* lichens are often leaflike. *Fruticose* lichens are often branched. Below: Patches of *crustose* lichens growing on an old wall.

Lichens are so hardy that they can grow almost everywhere, from the hottest deserts to the coldest polar regions. Large areas of the Arctic are covered with lichens known as reindeer moss because they are the main summer food of reindeer. Lichens cannot survive air pollution and they are rarely found in large cities.

Mosses and Liverworts

Most people can recognize mosses when they see them. The slender stems are no more than a few centimetres long and they carry thin, almost transparent leaves. There are no proper roots, although short hair-like outgrowths anchor the plants to the ground. Mosses usually grow in dense clusters, forming mats or cushions.

At certain times of the year dense clusters of leaves develop at the tops of the stems. Among these leaves there are the reproductive organs. The male organs are club-shaped and the female organs, which are usually on separate branches, are like tiny flasks. In damp weather, when the plants are covered with a film of moisture, the male cells are released. They swim in the film of water and the female 'flasks' produce a jelly which attracts the male cells. They then pair up with the egg cells in the flasks. The new cell formed by this union begins to grow and eventually forms a little oval box or capsule. This has its own stalk and it grows up above the rest of the moss plant. It contains the spores—minute dust-like particles that can grow into new moss plants. The spore capsule usually has a pointed cap to start with, but this comes off when the spores are ripe and reveals a complicated system of 'teeth'. When the air is dry these 'teeth' curl back and allow the breeze to scatter the spores.

When the moss spore germinates it sends out a thin green thread. This is called the *protonema* and it soon grows several branches. Buds develop on the branches and each bud grows into a new moss plant. As a result, the young mosses are clumped together right from the start. New threads and buds can grow from the base of the moss at any time, and so it can cover a large area of ground. Mosses grow best in moist, shady places.

Pellia showing the tough thallus and the simple spore capsules. Some liverworts, but not *Pellia*, grow buds in cups (below).

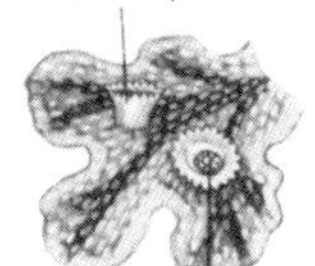

Liverworts

The liverworts are less well known than the mosses, although many of them are just as common. There are two main groups: flat and leafy. The flat liverworts grow in very damp places, such as river banks and woodland paths. The leafy liverworts are very much like the mosses, and are readily distinguished only when they are carrying spore capsules.

The life history of liverworts is very similar to that of the mosses. The spores are scattered by the wind and, when they reach suitable places, they grow into new liverworts. Many liverworts can also reproduce vegetatively by producing *gemmae*. These are tiny detachable buds produced in little cups.

Some common mosses of damp heathland.

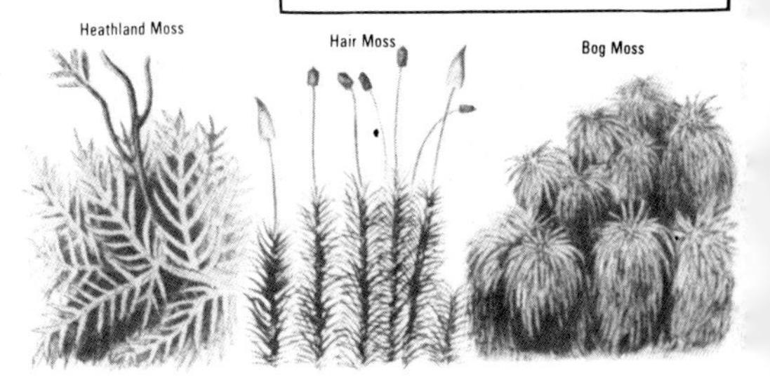

Ferns and Horsetails

The ferns and their relatives belong to a very ancient group of plants known as the Pteridophyta. They were in existence at least 400 million years ago and they included some of the earliest of all land plants. The pteridophytes reached their peak about 250 million years ago, when they formed great forests. Most of our coal consists of the remains of these plants, so we know quite a lot about them. Some of them reached heights of more than 30 metres. These huge plants have long since died out, and today's pteridophytes rarely exceed a metre or more in height.

Ferns and the other pteridophytes normally have roots and stems. Most also have leaves, but none of them ever has any flowers. They reproduce by scattering tiny spores, rather like those of the mosses and liverworts. The possession of proper roots is an advance over the lowly mosses, but even more important is the development of a system of tubes carrying water and food from one part of the plant to another. When the plants had evolved such a system they could begin to increase in size.

Ferns

Except in the tropical and subtropical tree ferns, which may reach a height of 20 metres, the stems are small and usually remain under the ground. They are normally unbranched, but the bracken has branching underground stems called rhizomes. These spread through the soil and throw up leaves.

The spores are normally produced on the undersides of the leaves, and in late summer you will see hundreds of little brown patches. These are patches of spore capsules. When they are ripe they burst open and the spores are scattered in the wind.

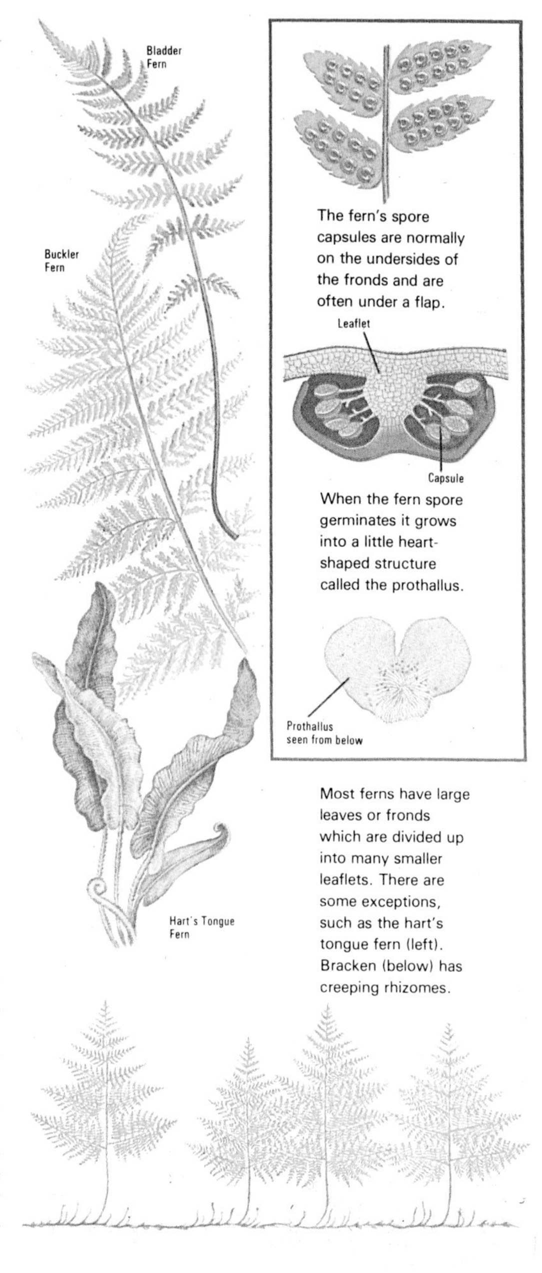

The fern's spore capsules are normally on the undersides of the fronds and are often under a flap.

When the fern spore germinates it grows into a little heart-shaped structure called the prothallus.

Most ferns have large leaves or fronds which are divided up into many smaller leaflets. There are some exceptions, such as the hart's tongue fern (left). Bracken (below) has creeping rhizomes.

When a spore falls on suitable ground it begins to grow, but it does not grow into a new fern plant. It forms a little heart-shaped plate up to 10 mm across. This plate is called the prothallus. It is green and has little root-like hairs on its underside. After a while the prothallus produces reproductive cells. The male cells are released in wet weather and they are attracted to the egg cells. They join up and the combined cell starts to grow into a new fern plant. It is supported by the prothallus at first, but it soon starts to make its own food. It may be several years before the young fern plant begins to produce its own spores.

Many fern plants will grow in dry places but, because the prothallus stage depends on damp conditions, ferns are found mainly in damp places.

Horsetails

About 250 million years ago the horsetails played a very prominent part in the world's plant life. Some of them were more than 30 metres tall. Today there are only about 25 species alive and they are rarely more than about 1 metre tall.

Horsetails are rather spiky plants without proper leaves. Some species have whorls of thin branches, but others have no branches and the stems look rather like long pencils sticking up. The stems are ridged and they bear a number of little 'collars'. These 'collars' are made up of tiny scale leaves and they occur at intervals all the way up the stem.

Underground, the horsetails have creeping rhizomes, like those of the bracken fern. The stems usually die down every year, but the rhizome lives on to produce the next year's stems.

The spores are produced in little 'cones' at the tips of the stems. Some horsetail species produce special spore-bearing stems which are brown and which come up before the green stems, but other species carry their 'cones' at the tips of ordinary stems. The spores grow into prothalli rather like those of the ferns, although they usually have upright 'fingers' on the upper surface.

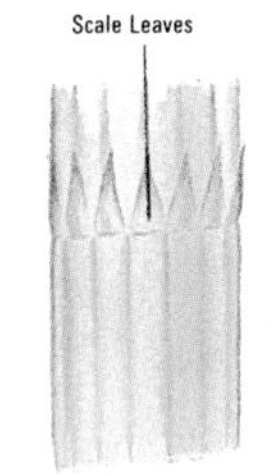

Horsetail stems are rigid and jointed. Each joint is surrounded by a collar of tiny scale leaves.

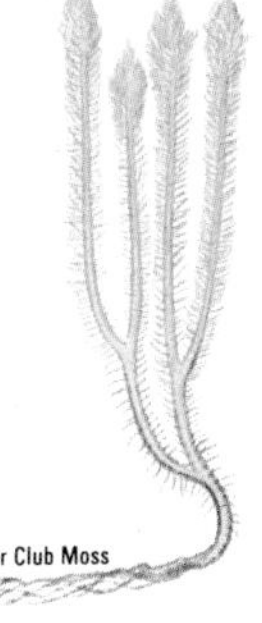

Club Mosses

Living club mosses are all quite small plants and, as their name suggests, they look rather like mosses. The spores are borne in 'cones' at the tips of upright shoots, and the cones are rather club-shaped—hence the common name. The prothallus of the club moss normally lives under the ground and gets its food by forming a partnership with a fungus.

There is a variety of ferns, especially in warm damp countries.

Conifers

Seeds are produced only by the two most advanced groups of plants—the cone-bearers and the flowering plants. All other plants reproduce by scattering tiny spores.

Apart from the strange, palm-like cycads, the cone-bearers are generally large trees. They are often called conifers. Examples include the pines, spruces (Christmas trees), cedars, larches, and monkey puzzle trees. Nearly all conifers have small narrow leaves called needles, and most of them are evergreens. The needles are very tough and coated with thick cuticles. The breathing pores are sunk well below the surface and water loss is therefore kept to a minimum. This is very important, because conifers generally live where water is often in short supply.

Conifers stretch in an unbroken belt from Norway to eastern Siberia and also right across Canada. The soil is frozen during the winter, so water is not available to the trees, but strong winds remove water by evaporation from the leaves. Conifers also flourish in the Mediterranean and other regions with moist winters and hot, dry

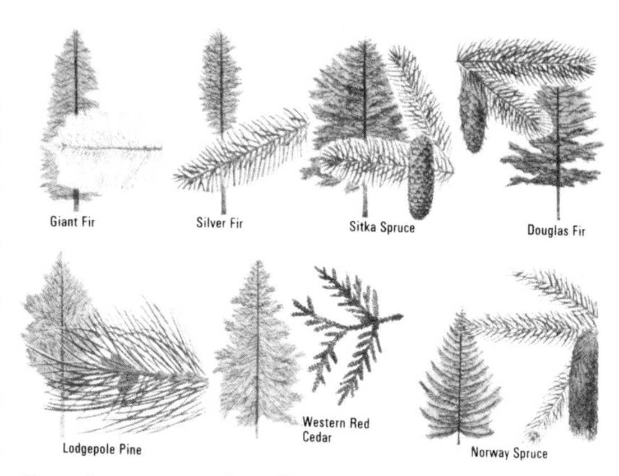

Above: Some well known conifers.

Coniferous forests cover vast areas in northern lands. They are our main source of timber.

Spores and Seeds

There is a big difference in the reproductive processes of the ferns and the conifers. The ferns produce spores, while the conifers produce seeds with baby plants in them.

Most ferns produce only one kind of spore, and all the prothalli (see page 23) are alike. In the conifers there are tiny male and female prothalli. They do not lead a separate existence.

Pollen grains are really the male spores. The pollen tubes and the few cells inside them are simply the male prothalli. The female spores are represented by cells inside the ovules. Only one spore forms in each ovule, and it grows into a bunch of cells representing the prothallus. One cell becomes the egg cell and is fertilized.

By retaining the female spore and prothallus inside the ovule, conifers avoid the need for free water. The early growth of the young plant takes place inside the parent, where it is protected and supplied with food.

Cycads

The cycads are strange plants part way between ferns and conifers. Most of them have short, stout trunks crowned with tufts of large fern-like leaves. The pollen and seeds are carried in large cones. When the pollen grain reaches the ovule it bursts and releases the male cell which swims to the egg cell.

A Scot's pine shoot

summers. The winter rains provide sufficient water for growth, and the needle-like leaves allow the trees to survive the summer droughts.

Coniferous trees grow more quickly than the broad-leaved or flowering trees and this is why conifers are planted so widely for timber. Their wood is softer and coniferous timbers are known as softwoods. Most of them are less durable than the hardwoods and they rot more quickly, although the western red cedar (not a true cedar) contains natural preservatives that make it resistant to rotting.

The Conifer's Life History

There are two types of cones—male cones and female cones—and they start to grow in the spring. Both types are normally found on one tree. The male cones are small and yellowish. They are usually rather inconspicuous, but they are easily seen in the pine because they are grouped into large clusters at the bases of the new shoots. The female cones develop near the tips.

Each cone consists of a series of overlapping scales. The scales of the male cone carry pollen sacs. Each female scale carries a pair of ovules which will eventually become the seeds. In dry weather the scales of the cones separate a little and pollen escapes from the male cones. Blown by the wind, the pollen grains enter the female cones. The cones then close up again, but the female cones begin to grow. They become green, then brown and woody. The pollen grains trapped inside the female cones send out fine tubes which grow into the ovules. A tiny cell from the pollen grain joins with the egg cell in the ovule to form an embryo plant. Food material is deposited around it and it gradually forms a root and a little shoot with several leaves. The whole thing is now a seed. The seed coat has a thin 'wing' growing out from one side and, when the cone opens, the seed can float away on the wind.

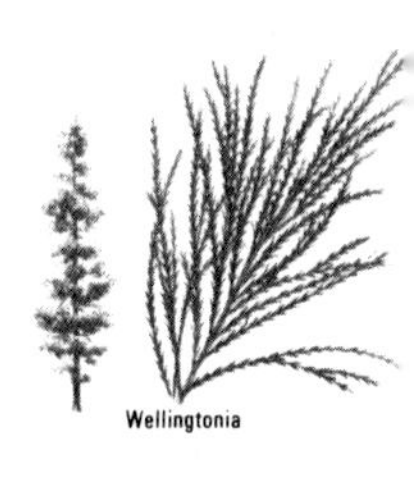
Wellingtonia

Spanish Fir

Scots Pine

Lawson's Cypress

Deodar

Flowering Plants

The flowering plants are the most advanced of all plants. Nearly every plant you see is a flowering plant. There are about 250,000 kinds.

Flowering plants range in size from the tiny floating duckweeds to huge forest trees. Their structure varies a great deal but, with the exception of a few parasitic and saprophytic species (see page 35), they all contain chlorophyll and they all make food by photosynthesis. Flowering plants also grow in much the same way as the non-flowering plants. It is their methods of reproduction that make the flowering plants so different from the mosses, ferns, and conifers.

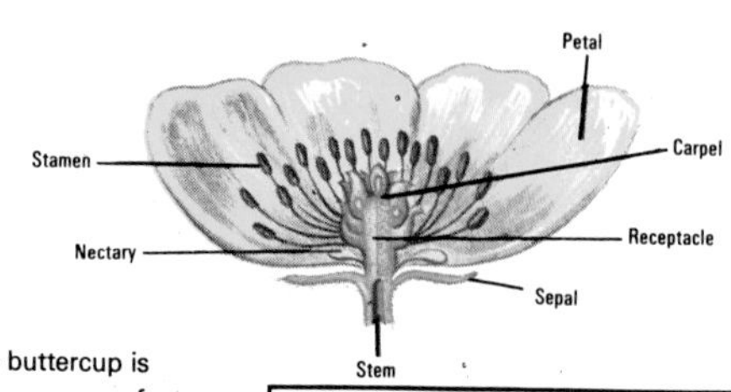

The buttercup is known as a perfect flower because it has all four kinds of floral organs.

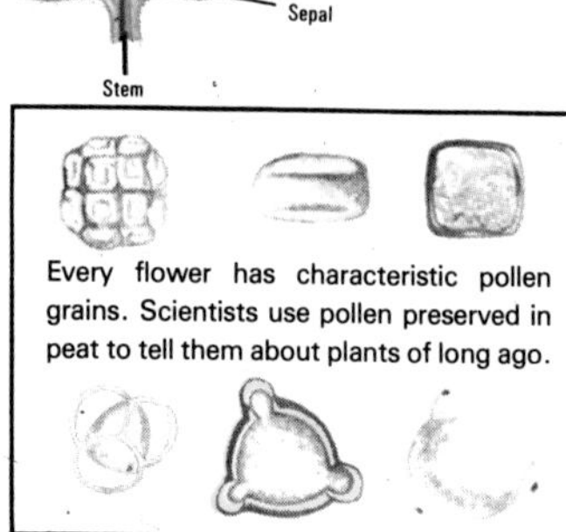

Every flower has characteristic pollen grains. Scientists use pollen preserved in peat to tell them about plants of long ago.

Most flowers are pollinated by insects. The insects are attracted by nectar, scent, and brightly coloured petals as in this cactus flower.

The Structure of the Flower

Conifers produce 'naked' seeds on the scales of their cones. Flowering plants, however, enclose their seeds in containers called fruits. Both the seeds and the fruits are produced by the flowers, and each part of the flower has a particular job to do.

Flowers are composed of very special leaves of which there are four main kinds. Around the outside of a flower there are usually a number of greenish leaves called sepals. In some flowers these are joined together. In others they are separate or free. When the flower is still in bud the sepals surround the other parts and protect them. When the flower opens the sepals bend back, or fall off. Just inside

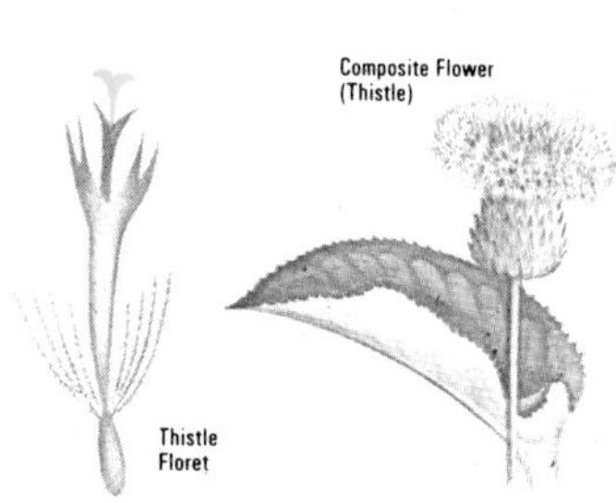

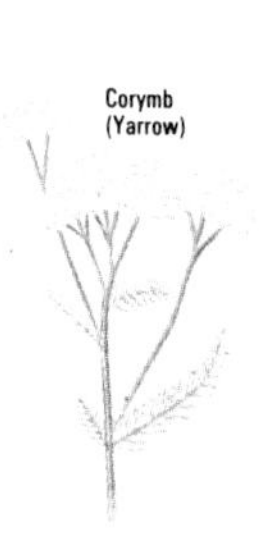

Right: Some plants, such as tulips, have only one flower at the top of the stem, but most have 'heads' of flowers with several flowers on one stem. The flower on a thistle or a dandelion plant is composed of hundreds of small flowers.

the sepals we find the petals. These vary a great deal in colour, shape and arrangement. They are often joined together. Inside the petals there are normally a number of pin-like objects. These are the stamens—the male parts of the flower which produce pollen. In the centre of the flower we find one or more carpels—the female parts of the flower. Each carpel contains one or more ovules which will eventually become seeds. The carpels may be joined together or they may be free. Each one has a sticky top called a stigma, which may be on a stalk.

Most flowers contain stamens and carpels, but there are some which contain only one or the other. Some species have male and female flowers on different plants, like willows and holly trees. Only the female trees bear fruit and seeds, so many holly trees never carry berries.

Pollination

Before a flower can form seeds it must be pollinated. This means that pollen of the right kind must fall on to the stigmas of the carpels. Some flowers are self-pollinated, that is pollen falls from the stamens to the stigmas of the same flower. Most flowers, however, have some way of preventing or discouraging self-pollination. The stamens may ripen and scatter their pollen before the stigmas are ready, or the stigmas may ripen before the stamens are ready. These flowers must be cross-pollinated with pollen from another flower.

Wind-pollinated flowers, such as these plantains, are generally drably coloured and they usually have long hanging stamens.

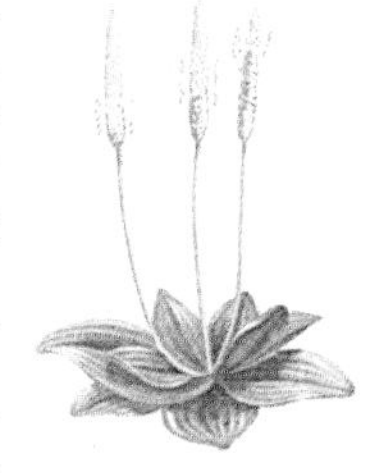

Some flowers can be pollinated only by certain insects. Only heavy insects can get into sweet peas.

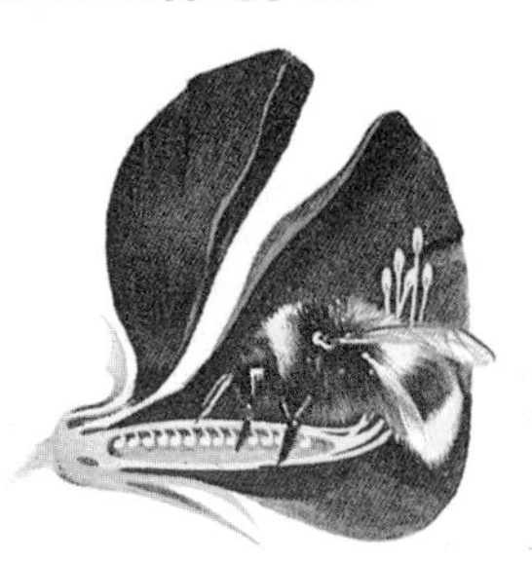

Some plants, including grasses and many trees, rely on the wind to carry their pollen. They generally have small and inconspicuous flowers, often without any petals. The stamens and carpels are often in separate flowers, and the male flowers often hang in long clusters called catkins. The stigmas of the wind-pollinated plants are usually branched and feathery for a better chance of trapping pollen.

The petals are the most obvious parts of the flower. They attract insects. After pollination, however, they wither and the carpels become obvious as they swell up and become fruits.

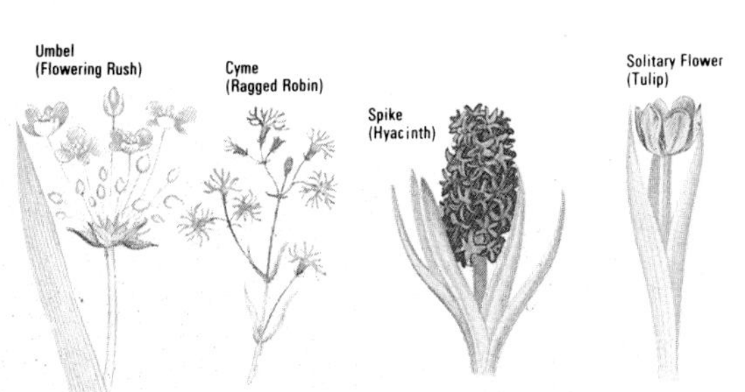

Most flowering plants rely on insects to carry their pollen from flower to flower. The insects are attracted to the flowers by the brightly coloured petals and by scent and sugary nectar. Most insects are interested only in nectar, and the flower's stamens and stigmas are cleverly placed so that the insects must brush against them as they reach for the nectar. The insects thus get dusted with pollen which they then carry to the stigmas of another flower.

rapidly grows into an embryo. This is a tiny plant, with root, shoot, and either one or two leaves. Food materials are deposited in or around these leaves, and the wall of the ovule becomes tough. It is now a seed.

While the seed or seeds have been growing, the carpel has been changing and now becomes the fruit. Fruits come in many shapes and sizes. Some are hard and woody, some are soft and juicy. All contain one or more seeds. Most fruits are formed from the carpel or carpels alone, but some are formed from other parts of the flower. Apples and pears, for example, are formed mainly from the flower stalk.

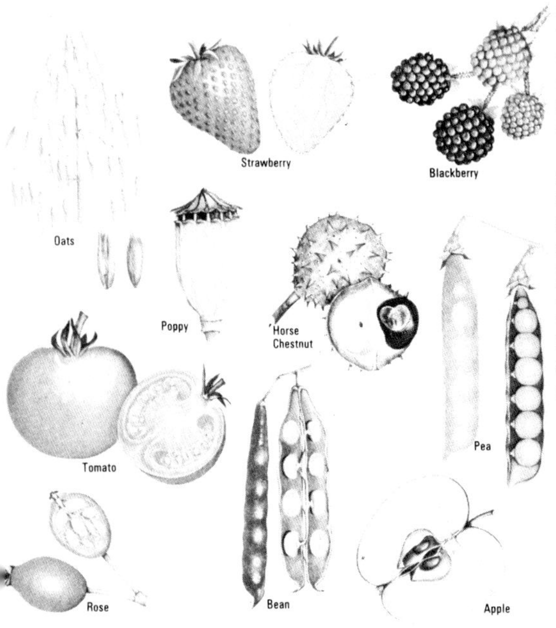

A selection of fruits. The apple, strawberry, and rose hip are false fruits.

The wind shakes out the seeds of poppy and snapdragon. Seeds of stock and violet fall out as the fruits dry.

The Formation of Fruits and Seeds

When a pollen grain falls on to a ripe stigma of the right kind it begins to grow. It sends out a thin tube into the carpel. Two minute cells from the pollen grain move down the tube and enter an ovule. One of them pairs up with the egg cell in the ovule. The egg cell is now said to be fertilized, and it

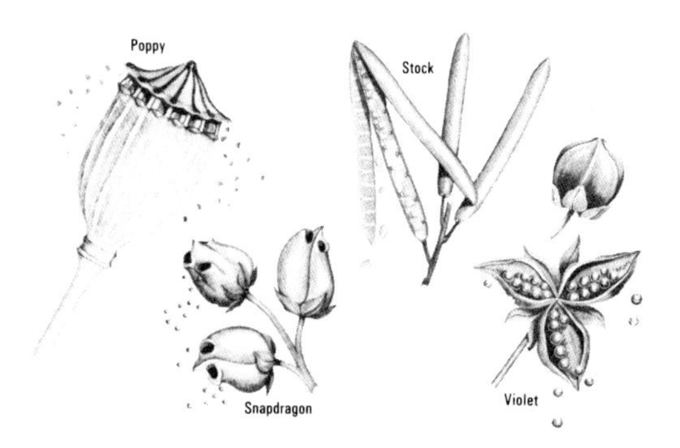

Many juicy fruits are eaten by birds. The seeds are then scattered far and wide.

Scattering Fruits and Seeds

When the seeds are ripe the plant must release them. Most plants have some method for ensuring that the seeds are carried some distance from the parent plant. The fruits play a large part in this scattering or dispersal of the seeds.

Some fruits, including the pods of peas and beans, dry out as they ripen. When completely ripe, the pods burst open and throw out their seeds. Many plants rely on the wind to scatter their seeds. Very small seeds are blown about by the wind quite easily. Larger fruits and seeds may have feathery outgrowths, such as the parachutes of dandelions and thistles. Many trees carry fruits which have wing-like flaps. When the fruits fall they are borne away on the wind.

Juicy fruits are attractive to birds and other animals, which eat the fleshy parts and spit out the seeds. The smaller seeds may pass right through the birds and come out unharmed with the droppings. Many fruits have small hooks which catch into the fur of animals.

Germination

When the seed has fallen on suitable ground, and when the temperature is high enough, it will begin to grow, or germinate. It will absorb water and swell up, and the seed coat will burst. Out will come the little root to grow downwards and anchor the young plant in the ground. The young shoot soon follows the root out of the seed and it grows upwards to the light. The seed leaves may grow up as well, although they often stay inside the seed. While the seed is germinating it relies on the food stored in it by the parent plant. Then, when its own shoot and leaves have grown up to the light, it starts to make food for itself.

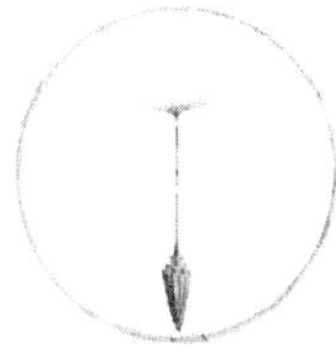

Composite flowers, such as thistles and dandelions, produce fruits with little 'parachutes' of hairs.

The fruit of maize and other grasses is a thin case around the starch-filled seed and is called a grain. Like all seeds, it contains enough food to last the seedling until it can make food for itself. First the root grows out from the seed and It begins to absorb water. Then the shoot grows up above the ground.

Annuals, Biennials and Perennials

Many plants live for only one season. They germinate as seeds in the spring, flower in the summer, and then die. They are called annual plants. Some of them actually have several generations in one year, taking only a few weeks to grow and produce their seeds. Most annual plants are rather small. Biennial plants take two seasons to complete their life histories. They grow from seed in the first season and store up supplies of food. These food stores keep them going through the winter and enable them to produce flowers and seeds in the following spring or summer. After flowering the plants die. Perennial plants go on living for several years—hundreds of years for some species of trees. Most of them flower many times during their lives. All trees are perennials, but so are some small plants such as dandelions.

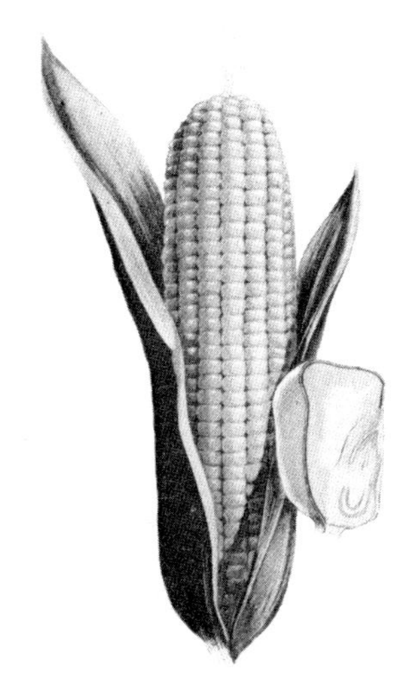

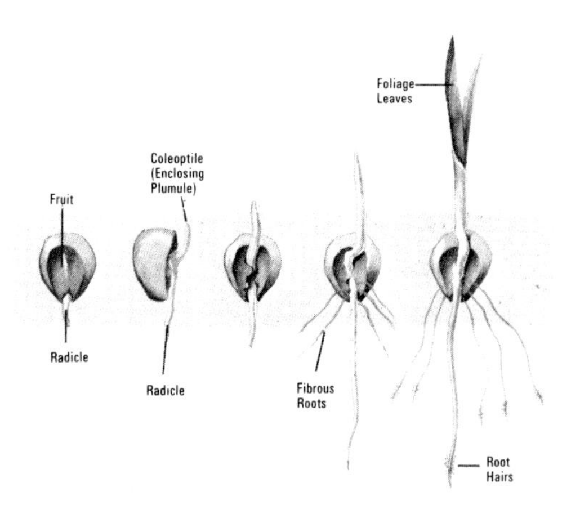

Flowering Trees

The main difference between trees and other plants is that trees are much taller and need strong woody trunks to support them. The trunk takes a long time to grow, however, and young trees are much like any other young plants when they first leave their seeds.

There are two main kinds of trees—the conifers and the flowering trees. The conifers carry their seeds in cones, and they are described on page 24.

Flowering trees belong to a wide range of families. The violet family, for example, contains many large tropical trees as well as the familiar violets. The pea family contains the herbaceous peas and clovers, the shrubby gorse and broom, and the attractive but poisonous laburnum tree. These plants all look very different, but you can see the connection if you look at the flowers.

Some flowering trees have very attractive flowers. Examples include apple, almond, lilac, horse chestnut, magnolia, and pussy willow. Many other trees have small and rather dull flowers which are easily overlooked. Examples of these trees include oak, ash, elm, maple, holly, and poplar.

Reproduction

Most inconspicuous tree flowers are pollinated by the wind, although the lime is a notable exception. Its hanging branches of yellowish flowers produce abundant scent and nectar. Wind-pollinated flowers usually open early in the year, often before the leaves have started to open. The hazel is a good example.

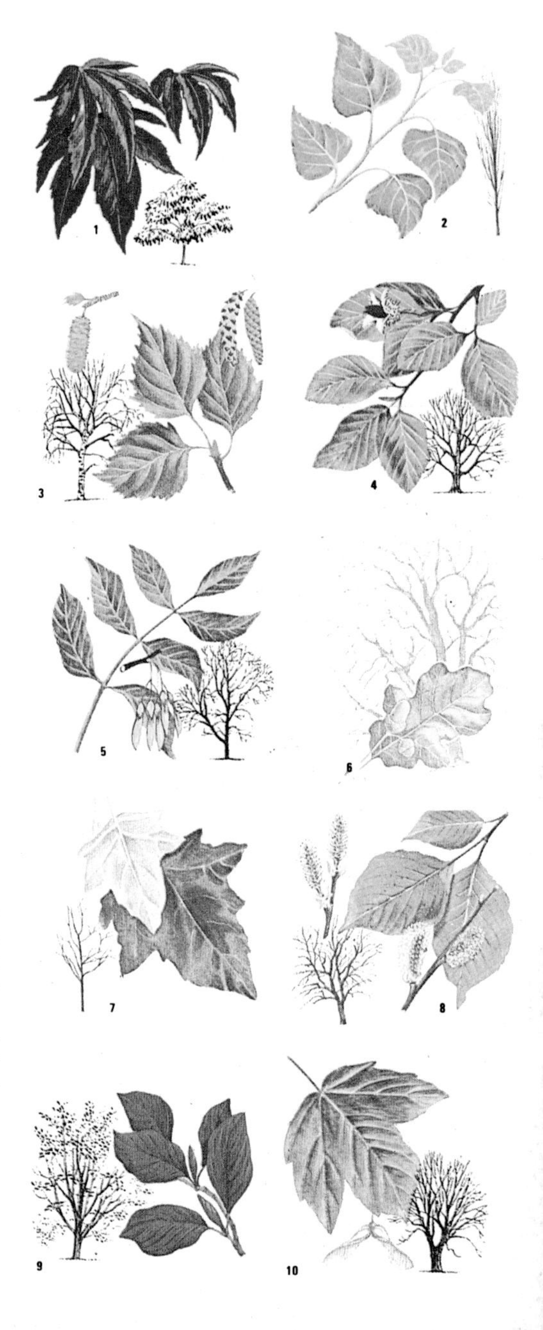

Right: A selection of flowering trees. 1. Japanese Maple, 2. Lombardy Poplar, 3. Silver Birch, 4. Beech, 5. Common Ash, 6. Oak, 7. White Poplar, 8. Goat Willow, 9. Purple Beech, 10. Sycamore.

All flowering trees carry their seeds in fruits of some kind or other. Many trees bear juicy fruits which are brightly coloured. These are eaten by birds, which then scatter the seeds over a wide area. Other trees may bear woody fruits. These may have 'wings' and drift away, or they may simply fall to the ground to be carried away by squirrels and other rodents.

Trees and Shrubs

Trees are woody plants which normally have one main stem or trunk carrying branches some way above the ground. The gorse, the gooseberry, and the rhododendron are clearly not herbaceous plants because they have woody stems, but they are hardly trees either. They have several stems of more or less the same size, all arising at about ground level. Plants which normally grow in this way are called shrubs.

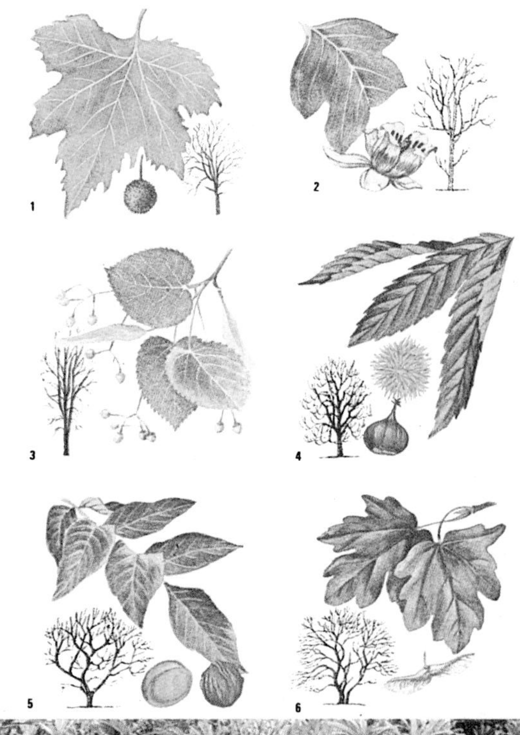

Right: 1. London Plane, 2. Tulip Tree, 3. Small leaved Lime, 4. Spanish Chestnut, 5. Walnut, 6. Maple.

Below: The attractive rhododendron is a typical shrub.

Grasses

The grass family, known as the Gramineae, is the most important of all plant families for man. From this one family come all our cereal crops: maize, wheat, oats, barley, rice, and many others. These cereals provide flour and they are the staple foods in most parts of the world. About two-thirds of the world's sugar comes from the tall grass that we call sugar cane. As well as feeding us, the grasses also feed all our important meat-producing animals.

If all the world's grass were to disappear we would lose a very large proportion of our food materials, and we would also lose our lawns, parks, and playing fields. Much of the land surface would become a barren waste because the grasses play a very important part in binding the soil together and preventing it from being blown away.

The grasses belong to the group of flowering plants called monocotyledons (see page 13) and, like the other members of the group, they have narrow leaves with parallel veins. They are nearly all small herbaceous (non-woody) plants, although some of the bamboos are very woody and reach heights of 30 metres.

The stems of most grasses are very short and close to the ground. Only the leaves grow upwards at first, and if you look at a lawn or a meadow in the spring you will see that it is nearly all leaf. This is what enables the grasses to stand up to repeated mowing or grazing. The stems are not removed and they can easily send up a new crop of leaves. The stems also send out a number of side shoots at ground level. These are called tillers and they enable the grass to cover the ground and form a turf quite quickly. If the main stem is damaged or removed the plant will produce even more tillers, and so mowing, trampling, and grazing actually help to form a thick turf.

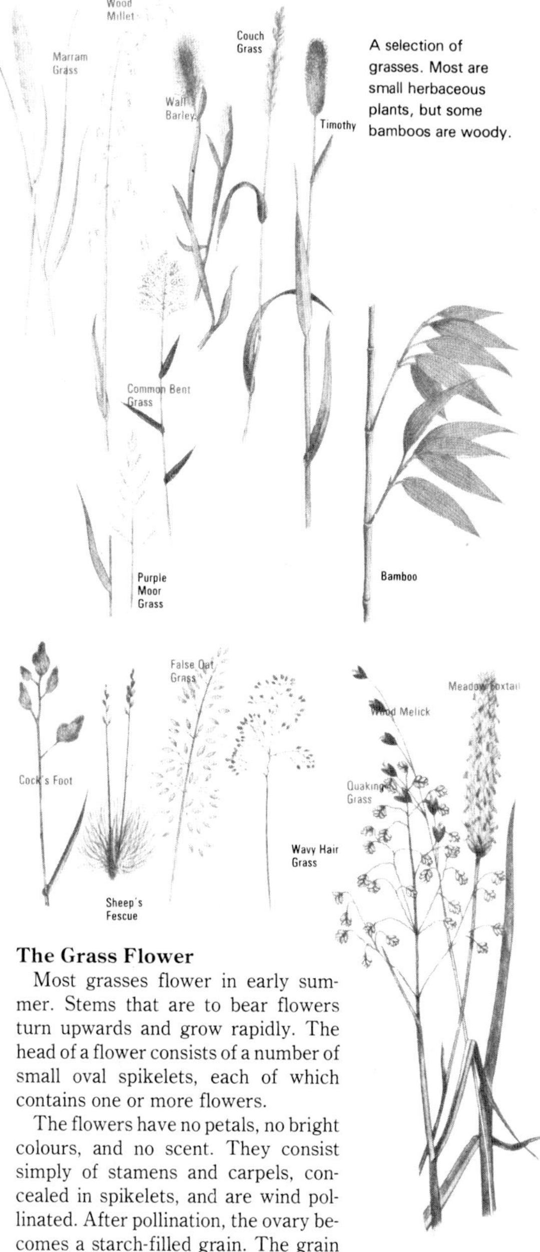

A selection of grasses. Most are small herbaceous plants, but some bamboos are woody.

The Grass Flower

Most grasses flower in early summer. Stems that are to bear flowers turn upwards and grow rapidly. The head of a flower consists of a number of small oval spikelets, each of which contains one or more flowers.

The flowers have no petals, no bright colours, and no scent. They consist simply of stamens and carpels, concealed in spikelets, and are wind pollinated. After pollination, the ovary becomes a starch-filled grain. The grain of grasses is important as human food.

Climbing Plants

Tall plants need sturdy stems to support them and this means that they grow rather slowly as a rule. But some plants have overcome this problem by climbing over their neighbours. These climbing plants do not need strong stems, so they can put all their efforts into upward growth and grow very quickly. Climbing plants have developed many methods of climbing.

The young stems of ivy creep over the ground and when they find a wall or a tree trunk they turn upwards. Small roots grow from the stems and take a hold in cracks and pores. They hold the slender ivy stems tightly against the support. The ivy does not take food from the tree on which it grows, but a lot of ivy can weaken a tree.

Many plants climb by simply twisting or twining themselves around their neighbours. Honeysuckle and bindweed are common examples, and so are the runner beans. Some species twine clockwise and others twine anticlockwise.

Peas and several other plants possess slender outgrowths called *tendrils*. They are often specially modified leaves. They coil tightly around any twig they happen to touch and hold the climber firmly in position.

Right: A selection of climbing plants. Clematis climbs by twining its leaf stalks (petioles) about supports. Honeysuckle is the best known of the 'clockwise' twiners and bindweed one of the most common 'anticlockwise' climbers. Ivy climbs by means of aerial roots. The tendrils of the Virginia Creeper may develop into adhesive discs which enable the plant to climb.

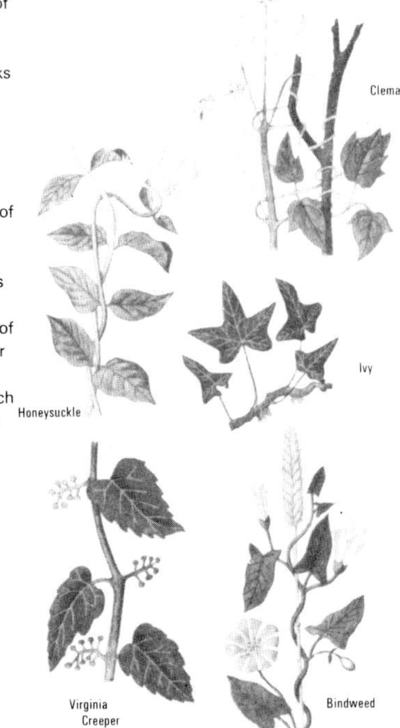

The prickles of roses or blackberries are not basically for the protection of the plant. They are climbing aids and they hook over the twigs of other plants to support the climber.

Below: Ivy is a common climber of both trees and walls. Its aerial roots secrete a sticky fluid which glues them to the support.

Insect-Eating Plants

A number of plants supplement their food supplies by catching and digesting insects. The insects provide additional nitrates and other minerals, so plants with insect-catching habits are often able to survive on poor soils. Most of the insect-eating plants actually live in peat bogs and similar wet places. There are relatively few soil bacteria in such places so the plants do not decay very rapidly and very little mineral matter becomes available for growing plants.

Insects are caught by specially modified leaves or parts of leaves. They are then digested with juices rather like those in our own intestines. All insect-eating plants can survive without catching insects. If grown in good soil they will grow and flower quite normally. When grown on their normal soil, however, they need the extra minerals and they do not flower well if they are unable to catch insects.

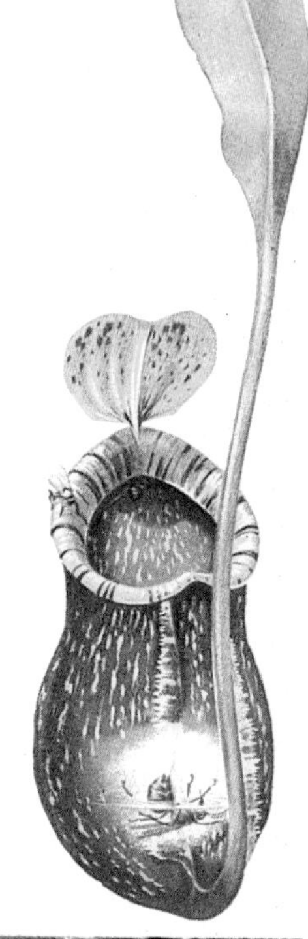

Pitcher plants trap insects in various kinds of 'jars' which are formed from the leaves. The 'jars' contain fluids which attract the insects and then digest them when they fall in. The digested materials are absorbed by the plant. Some pitcher plants have lids on their pitchers to prevent rain from diluting the juices.

The sundew attracts insects with the shiny droplets on its leaf. The insects are trapped by the sticky fluid and the leaf closes up.

Parasitic Plants

Mistletoe is best known as a Christmas decoration, but it is interesting because of its strange way of life. It has no roots and it grows on the branches of other trees, especially apples and poplars. It sends suckers into the branches to take in water and mineral salts. The mistletoe is therefore a parasite—a 'lazy' organism that gets its food from another plant or animal. The mistletoe is only a partial parasite, because it can make its own food.

Another very common partial parasite is eyebright. It has roots of its own, but they are not very efficient and they obtain most of their water and minerals by sending suckers into neighbouring grass roots. Eyebright can grow without the help of the grasses, but it is stunted and does not flower well.

A number of flowering plants have become complete parasites. They have no chlorophyll so they cannot make food for themselves: they take everything they need from other plants. The only thing these parasites have to do for themselves is produce flowers and seeds. All their energies are concentrated in this direction and all other parts of the plant are reduced.

Dodder is a well known parasite whose tangled red stems often smother low growing gorse and heather in the summer. The thread-like stems twine around the host stems and send suckers into them. The host plants often become discoloured and they may die off. The dodder's leaves are reduced to tiny scales and it does not even have a root after the first few days of life.

The World's Biggest Flower

Rafflesia, which grows in Malaysia and neighbouring countries, has no roots, stems, or leaves. It is a parasite of vines and produces a flower 60 centimetres across. A seed becomes wedged in the bark of a vine and sends threads into its wood. The threads take food from the host and later produce a flower bud at the surface.

Looking like pink strands of cotton, the dodder twines around a heather plant and sends suckers into it to obtain food.

Protection in Plants

Some plants are unpleasant to touch. They have sharp thorns or prickles capable of inflicting painful wounds, or possess poisonous hairs that produce irritating rashes.

Thorns and Spines

The hawthorn and the blackthorn or sloe tree both have very unpleasant prickles, which you will probably have felt if you have tried to pick their attractive flowers or fruits. These prickles are actually dwarf branches. They develop from buds, but they soon stop growing and become sharply pointed. Prickles of this kind should really be called *thorns*. We can see that they are formed from special stems because they often carry leaves.

In some other plants the leaves form spines. The barberry has some of its leaves modified as long, slender spines. Gooseberry and acacia spines are outgrowths from the leaf bases. Gorse bushes have both thorns and spines: the leaves are needle-like and short, pointed stems grow in their axils. Another form of spine develops when the edges of leaves are drawn out into sharp points, as in holly and thistles. The spines of cacti are modified leaves. They always arise on little 'cushions' called *areoles*. This will always distinguish a cactus from any similar prickly plant.

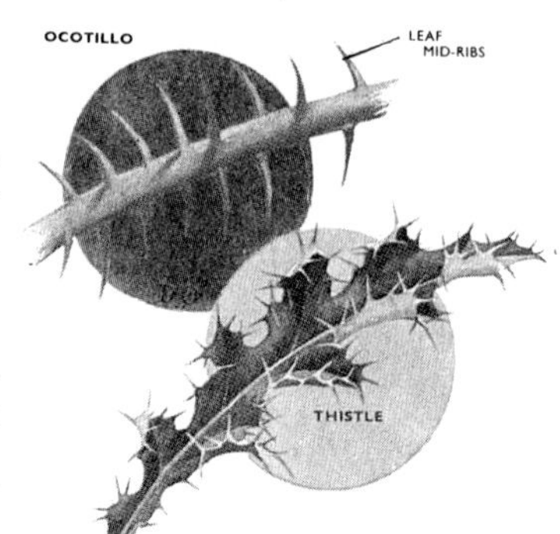

The ocotillo is a desert plant of America. Its spines are modified leaf midribs. Leaves appear after rains but are quickly dropped again in the dry season, leaving the pointed midribs.

The spines of the thistle are formed from the stiffened leaf margins.

The prickles of the blackberry are climbing aids. The spines of cacti are rather a puzzle. They probably have several functions, including the trapping of a layer of still air around the plants. This helps to cut down evaporation.

The curved prickles of roses and blackberries are outgrowths from the skin. They are often called *emergences*. Their main job is to help the plants climb (see page 33).

The Function of Prickles

Most prickly plants live in dry or semi-dry regions, or else in places where water drains readily from the soil. The plants must not lose too much water by evaporation, and one of the commonest ways of cutting down

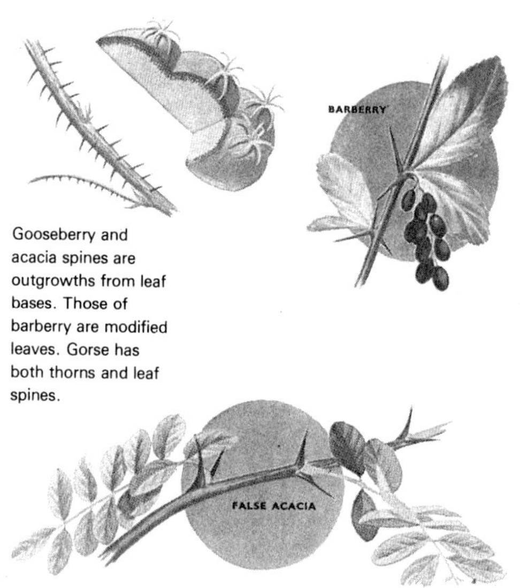

Gooseberry and acacia spines are outgrowths from leaf bases. Those of barberry are modified leaves. Gorse has both thorns and leaf spines.

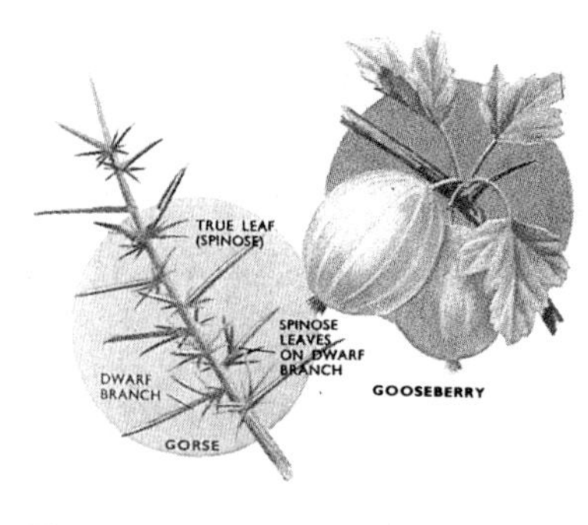

water loss is by reducing the size of the leaves. Prickles seem to be mainly concerned with reduction of water loss. Some spines, however, are probably basically for protection. Holly has very spiny leaves on its lower branches—where animals can browse—but its higher leaves are often quite smooth.

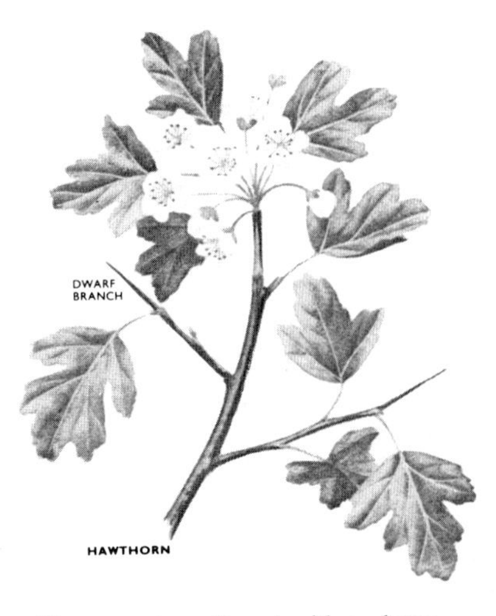

The thorns of the hawthorn are dwarf branches, shown by the fact that they often carry leaves.

Stinging Hairs

Most plants have some sort of hair on their stems and leaves. Long coats of hair may keep a plant warm in a cold climate, or they may keep the sun off it in a hot climate. The hairs also help to cut down the loss of water. Goosegrass or cleavers has tiny hooked hairs which help it to climb.

The stinging hairs of the stinging nettle grow on the stem and leaves. They are strengthened with a glassy material called silica and they end in little spherical swellings. These break off when something brushes against the nettle and the sharp needle-like point sticks in. Poison then flows into the wound from the cavity of the hair and it sets up an irritating rash. The poison is a complicated mixture of substances, with no real antidote. The fever nettle of West Africa has a very strong poison that can make a man quite ill. Even elephants are said to avoid this plant.

Left: Blackberry fruit, with the curved prickles also visible on the stem.

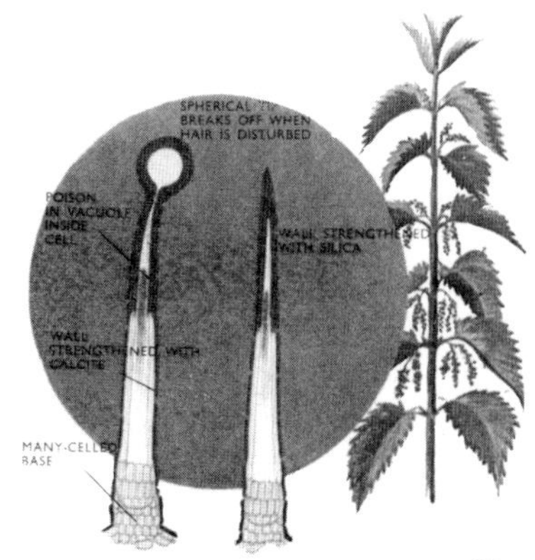

The stinging hair of the nettle ends in a little round swelling. This breaks off when the plant is touched, leaving a sharp needle which injects poison into your skin.

The Soil

> **Climate and Soil**
>
> Although the nature of the underlying rock has a very great effect on the nature of the soil, the climate has the final say as long as the ground remains undisturbed. In the cold temperate regions, the rainfall exceeds the evaporation of water from the soil. Humus and minerals such as iron and calcium are therefore continually washed down from the surface layers. This action, known as leaching, is especially obvious in sandy soils and the surface layers become quite pale. Leached soils are called *podsols*.
>
> In the warm temperate regions the evaporation of water more or less equals the rainfall and so there is not much downward movement of humus and minerals. The surface layers are brown and the soils are called *brown earths*. *Black earths* are found under the grass of the steppes and prairies. These regions are fairly dry and water is nearly always moving up to the surface. Minerals and humus remain at the surface and the surface layers are black.

The soil is a very complex material that is formed mainly from the underlying rocks. You can see the soil-forming processes at work in old quarries and in road or railway cuttings. Wind and rain attack the bare surfaces and small pieces begin to crumble away. These little rock particles form the soil skeleton. Small grasses and other plants can take root among them and the roots find their way into tiny cracks in the rock, thus helping to break it up further. When the plants die their remains decay and add organic matter called humus to the soil skeleton. A true soil is now beginning to form and larger plants can take root in it. The underlying rock is broken down further and the soil gets deeper. In time it becomes a mature soil, with plenty of humus. A soil also contains air spaces, water, and hordes of tiny organisms which help to break down dead plants and animals.

Because the soil is composed mainly of rock particles, different types of rocks produce different kinds of soils. Soft clay rocks provide very tiny mineral particles which hold a lot of water and which stick together very readily. Such particles produce wet sticky soils. Sandy rocks, on the other hand, produce large particles and light, well-drained soils.

These different types of soil support very different types of vegetation. Sticky clays, for example, support oak woods. Sandy soils support pine woods or heathland. Chalk and limestone soils contain a great deal of lime.

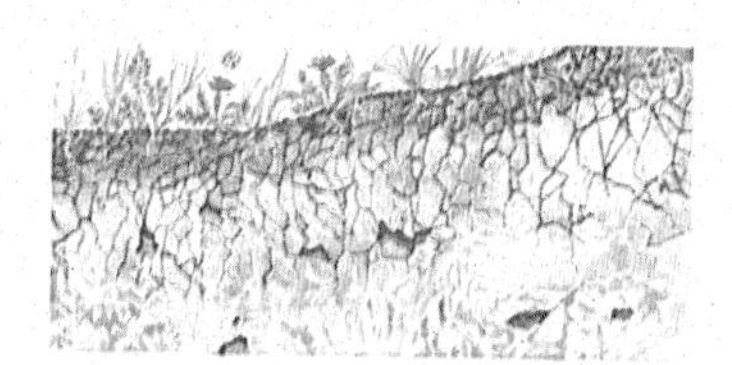

Chalk and limestone soils are rather thin, but they carry a wealth of attractive lime-loving plants.

The chalk downs of England are famous for their orchids, although there are not nearly so many now as there were 100 years ago. The fragrant and pyramidal orchids, seen here with yellow bird's foot trefoil, are among the commonest species.

Roots and Stems

If you have ever tried to pull up a tree seedling, or even a dandelion, you will know that it is hard work. The root holds the plant very firmly in the ground. This is one of the main jobs of the root. It anchors the plant so that the wind cannot pull it out of the ground. The other job of the root is to absorb water.

Dandelions, thistles, and many other plants have a *tap root*. This is a single main root which grows more or less straight down into the ground with only a few branches. Not all plants have tap roots. If you pull up a grass or a daisy plant you will see that it has lots of roots, all more or less the same size. These roots are called *fibrous roots*.

The root anchors the plant by growing down between the soil particles and filling the spaces very tightly. Both types of roots also absorb water from the soil. The finer regions of the roots, a little way behind the root tip, bear thousands of *root hairs*. These are very tiny and they are usually broken off when a plant is uprooted. The root hairs have such thin walls, water can seep into them from the soil. This water enters the long conducting tubes and begins its journey up to the leaves. The conducting tubes have very tough walls and give the root its strength.

Roots hairs grow out from the outer cells of the root and greatly increase its surface area and the amount of water that the root can absorb.

Old Root Hairs
Vacuole
Root Hairs at Maximum Length
New Root Hairs
Growing Tip

The conducting tubes in a young stem are collected into little bundles. The outer part of each bundle consists of numerous tough fibres which support the stem.

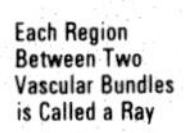

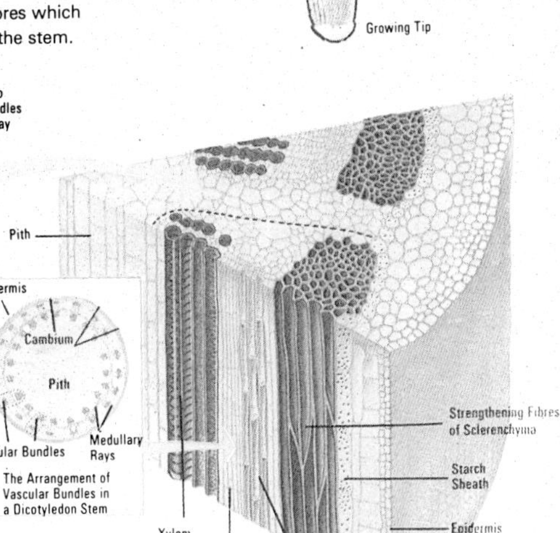

The Arrangement of Vascular Bundles in a Dicotyledon Stem

A tree trunk consists almost entirely of tough fibres and water-carrying tubes. Food is carried only by tubes in the bark.

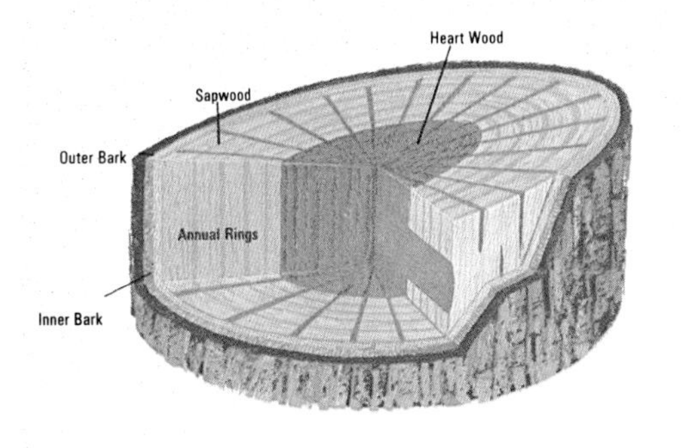

The Stem

The stem of a plant has to carry the water from the roots to the leaves. Stems are usually upright and they may or may not have branches. Some stems are quite soft and juicy. They are called *herbaceous* stems. Other stems are woody.

The conducting tubes are continuous with those in the roots, although the arrangement is different. The tubes of a young stem are arranged in little clusters called *vascular bundles*. They form a ring in the outer half of the stem. The tough walls of the tubes, together with the long fibres that

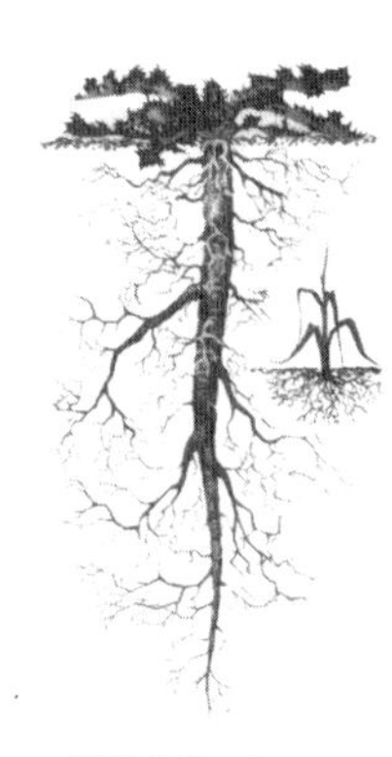

A tap root system, shown by the thistle, has one main root with several smaller branches. A fibrous root system consists of many similar roots.

surround them, give great strength to the stem yet they still allow it to bend in the wind. As well as the water-carrying tubes, the vascular bundles contain food-carrying tubes.

As a stem gets older its vascular bundles join up to form a complete ring of tubes and fibres. Woody plants form more tubes and fibres each year and the trunk gets thicker. It consists almost entirely of tubes and fibres. The older ones get squashed in the middle of the trunk and they form the *heart-wood*. They no longer carry water. The outer regions, which do carry water, form the *sapwood*.

Annual Rings

In those parts of the world where there are definite seasons, the trees slacken their growth in the autumn. Only narrow conducting tubes are formed at this time. No new tubes are formed in the winter, but large ones are formed when the trees 'wake up' in the spring. There is a distinct boundary between the autumn wood and the spring wood and so, when a tree is cut down, we can see the *annual rings*. Counting the rings shows how old the tree is.

Why a Plant Needs Water

All living things need water because protoplasm is nearly all water. Water is necessary to dissolve food materials so that they can be absorbed from the soil and carried about in the plant. Water is also one of the basic raw materials from which plants make sugar and starch (see page 41). Plants also need water for support. This is very important in herbaceous plants which do not have many woody fibres to support them. The water makes the cells firm, or turgid. When the cells of a plant are turgid the whole plant stands up, but if the water is lost the cells collapse and the plant droops.

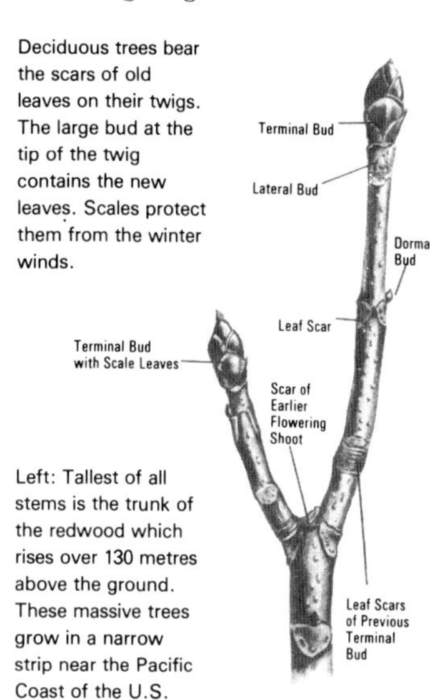

Deciduous trees bear the scars of old leaves on their twigs. The large bud at the tip of the twig contains the new leaves. Scales protect them from the winter winds.

Left: Tallest of all stems is the trunk of the redwood which rises over 130 metres above the ground. These massive trees grow in a narrow strip near the Pacific Coast of the U.S.

How Plants Make Food

The sun provides almost all of the energy on the earth, including all the energy that we use in our bodies to move about and to keep warm. But our bodies cannot harness the sun's energy directly, and nor can any other animal. This can be done only by green plants. They make use of the sun's light energy and they convert some of it into chemical energy which they store in their bodies. If it were not for the plants, animals would be unable to survive, because only the plants can make energy-containing foods from the simple materials in the air and the soil.

Photosynthesis

The basic source of all food is the remarkable process called photosynthesis, which takes place in all green plants. Photosynthesis means 'building with light' and it is the process by which green plants make sugars from carbon dioxide and water. It takes place only in the presence of sunlight and in the presence of the green colouring matter called chlorophyll.

The chlorophyll absorbs some of the light energy falling on it and it uses it to power a complicated chain of reactions involving water and carbon dioxide. Oxygen is given off in the process, but the chlorophyll is unchanged and goes on absorbing more light and passing it on to more water and carbon dioxide.

Leaves

Most of the photosynthesis in an ordinary plant takes place in the leaves. They are specially constructed for this work. Most leaves sit horizontally on the plant and the stems ensure that the leaves are arranged so as to catch the greatest amount of light. If you look up from the base of a tree, you will see how efficiently the leaves trap the light.

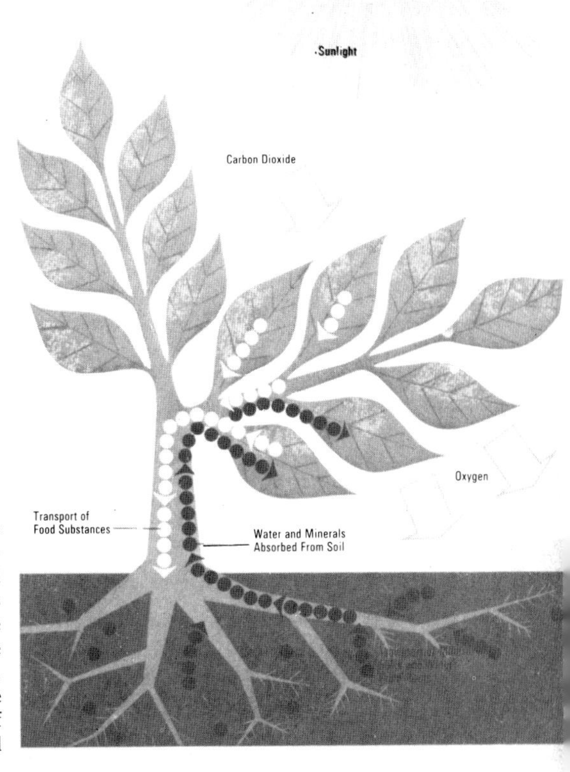

Carbon dioxide enters the plant through the leaves. Water enters through the roots. They combine in the leaves to form sugar, which is then carried away to all other parts of the plant. Oxygen is given out during the process and it escapes through the leaves. The whole process is powered by sunlight.

The water needed for photosynthesis comes from the roots and enters the leaves through the veins. The carbon dioxide comes from the air and enters the leaves through tiny breathing pores called stomata. Most of these are on the lower surface of the leaves. As well as letting carbon dioxide into the leaf, they let the unwanted oxygen out. They also let the excess water vapour out, but if water is in short supply they will close up and prevent the plant from losing water.

The cells inside the leaf are always surrounded by a film of water and the carbon dioxide dissolves in this film. It can then pass through the cell walls and into the protoplasm where the chlorophyll is. The protoplasm

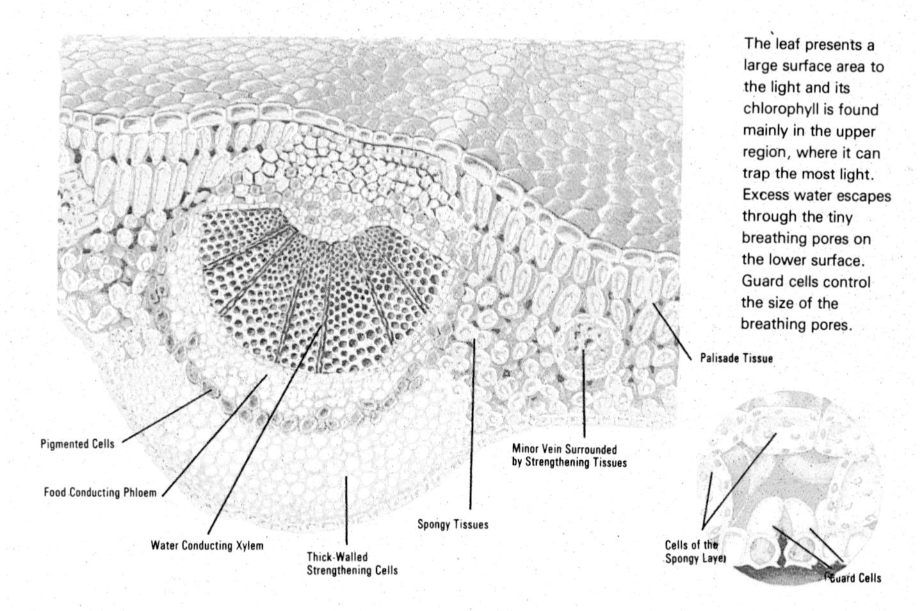

The leaf presents a large surface area to the light and its chlorophyll is found mainly in the upper region, where it can trap the most light. Excess water escapes through the tiny breathing pores on the lower surface. Guard cells control the size of the breathing pores.

consists largely of water and so, as long as the sun is shining, the plant can carry out photosynthesis.

Plants which grow in shady places often make up for the low light intensity by having more chlorophyll in their leaves. This makes them somewhat darker than plants growing in full light.

After Photosynthesis

The process of photosynthesis provides the plants with sugars. These can be used to provide energy. They can also be converted into starch and stored. Some of the sugars are converted into cellulose and used to build up more plant material. But the plant cannot exist just on the sugars and starches. It must make more protoplasm if it is to make more cells and grow. Protoplasm contains proteins, and proteins contain nitrogen and various other materials called minerals. The plant must obtain these minerals from the soil. They are combined with sugars to make proteins.

Mineral Nutrition

The plant body consists mainly of the four elements hydrogen, carbon, oxygen, and nitrogen. The first three elements are obtained from water and carbon dioxide and they are always freely available. There is plenty of nitrogen in the air as well, but plants cannot use this atmospheric nitrogen. They have to absorb nitrates or other nitrogen-containing compounds from the soil.

Nitrogen is essential for the growth of plants because nitrogen is a constituent of all proteins, and proteins are the basis of all living matter. If a plant is not able to get enough nitrogen it will not be able to grow.

Among the other elements in a plant's tissue, there will be iron, magnesium, calcium, potassium, sodium, silicon, phosphorus, sulphur, and chlorine. The amounts vary a great deal, and some of the minerals may not be essential—they are there in the plant simply because the plant had absorbed them in water taken from the soil. Apart from hydrogen, carbon, oxygen, and nitrogen, the main elements or minerals needed by plants are: magnesium, iron, calcium, potassium, phosphorus, and sulphur. A lack of any of these in the soil will interfere with the proper growth and development of the plant.

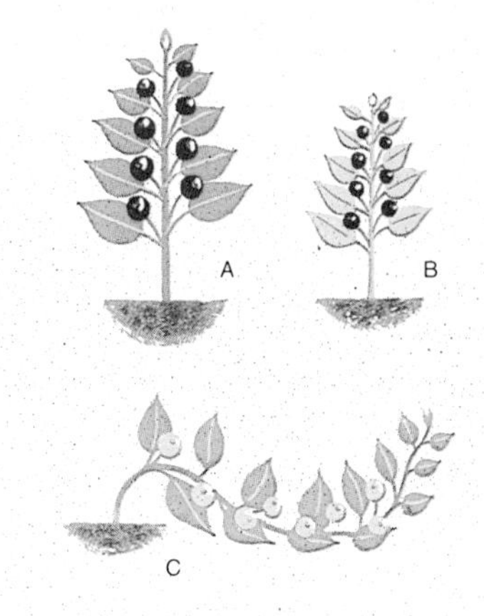

Right: The effect of nitrogen on plant growth.
A. Nitrogen encourages rapid leaf growth.
B. Insufficient nitrogen means poor, weak growth.
C. Excessive nitrogen can cause soft rank growth.

Below: Spraying trace elements on pasture land in New Zealand. Large areas have been made more fertile by the addition of small amounts of manganese, molybdenum, zinc, and copper.

In nature, plants grow and die and their remains decay on the ground. The minerals are then returned to the soil for a new generation of plants. Animals also leave droppings which decay and release minerals. The farmer and the gardener, however, do not leave their crops to rot on the land. They take them away to sell or to eat. Minerals are not returned to the soil and so the farmer and the gardener have to add extra minerals every now and then in the form of manure or fertilizer. Most soils contain plenty of calcium and they also have enough iron and sulphur for the plants' needs. The only minerals that normally have to be added are nitrogen, phosphorus, and potassium. Most fertilizers provide one or more of these elements.

Food Storage in Plants

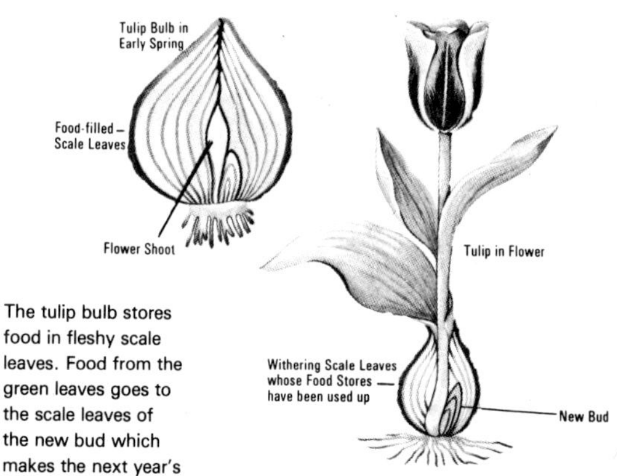

The tulip bulb stores food in fleshy scale leaves. Food from the green leaves goes to the scale leaves of the new bud which makes the next year's bulb. The old scale leaves remain as thin brown coating scales.

Plants do not need a lot of energy and much of the food they make is stored for later use. It may be stored up to tide the plant over the winter and to get it going again in the spring, or it may be stored up in readiness for flowering and the production of fruits and seeds.

During the process of photosynthesis plants make sugar, but most of them convert the sugar to starch before they store it. Proteins and fats or oils are stored by some plants, especially in seeds.

It is very lucky for mankind that plants do accumulate stores of food materials because we can take the food for ourselves. Our vegetable crops have all come from wild plants and most of them have been cultivated for a very long time. During that time, the size of the food-storing part of the plant has been greatly increased and cultivated plants give a much greater yield than their wild cousins.

Storage in Roots

Many plants store their food in swollen roots. Two well known examples are the carrot and the sugar beet. Each plant has one main root, the tap root. These plants, and most of the others like them, are biennials. During the spring and summer of the first year the plants make food and store it in their swollen tap roots. The plants die down in the autumn and the food reserve in the roots is used to produce new leaves, flowers, and seeds in the spring. The plants then die.

Other plants storing food in their roots include the dahlia and the lesser celandine. They store food in swollen roots called root tubers.

A corm is a swollen stem. Buds develop on the corm which forms each year at the base of the flower shoot.

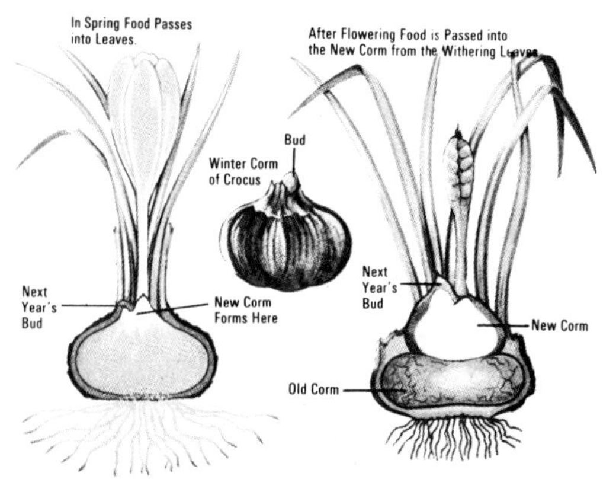

Storage in Stems

Many plants store food in their stems without any special modifications. Others, however, develop specially swollen stems. Kohl rabi, a member of the cabbage family, stores food in a tennis ball-sized swelling just above the ground. Irises and several other plants possess horizontal underground stems called rhizomes. Potato tubers are also underground stems used for storing food.

Corms are underground stems, although they are often called bulbs. The crocus corm is a short, rounded stem with one or more buds on the top.

Leaves and Flowers

Cabbages are some of the best known plants that store food in their leaves. The ordinary cabbage consists of a tightly packed bunch of leaves, all full of food. Bulbs also store food in leaves, although these are rather special leaves. A bulb consists almost entirely of swollen, juicy leaves called scales. These surround one or more buds. When the bulb starts to grow the food is passed from the scales to the buds. The scales then wither, but a new set of scales is formed inside them. They will be the next year's bulb. The papery coverings are the shrivelled old scales.

Some plants even store food in their flowers. Best known of these is the cauliflower. The large creamy heads that we eat are the young flowers, tightly packed together because their stems have not grown.

Fruits and Seeds

All seeds must have some sort of food reserve for the seedlings to grow to the stage at which they can start making their own food. Many of the larger seeds—peas, beans, cereal grains, nuts, and so on—are used for human food. Many fruits also contain food materials. Sweet, juicy fruits attract animals but, while eating the fruits, the animals scatter the seeds. This is of great advantage to the plants and it is worth their while to provide food for the animals.

Above right: Collecting sugar beet for processing. The swollen tap roots contain the plants' food reserves, designed to produce new leaves, flowers, and seeds in the spring.

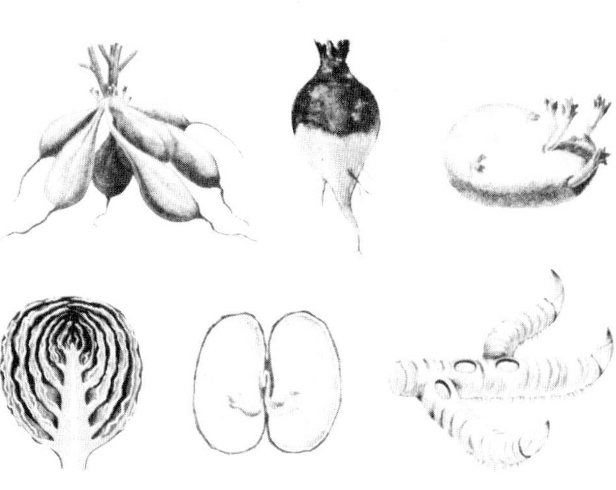

Right: A selection of plant food stores including roots (dahlia tubers and sugar beet), stems (potato tuber and rhizome), seeds (bean seed), and leaves (cabbage).

Plant Growth and Movement

The growth pattern of a plant is flexible. A tree growing in the middle of a wood has a very different shape from the same kind of tree growing by itself in the middle of a field. Plants also differ from many animals in that they go on growing throughout their lives.

Stems and Roots

The buds of plants are very short stems surrounded by young leaves. The leaves on the outside protect the bud during the winter. The bases of all the leaves are very close together. In other words, the *internodes* of the stem —the regions between the leaf bases— are very short. In spring the leaf scales fall from the outside of the bud and the internodes lengthen. The new leaves separate and begin to unfold. The actual tip of the stem is never seen because it is always sheathed by developing leaves. Later in the summer more scale leaves are produced around the stem tip and growth slows down. The scales protect the stem tip and its leaves during the next winter.

Many of the subsidiary buds on a stem never open. They are reserve buds, which develop only if the main stem is damaged.

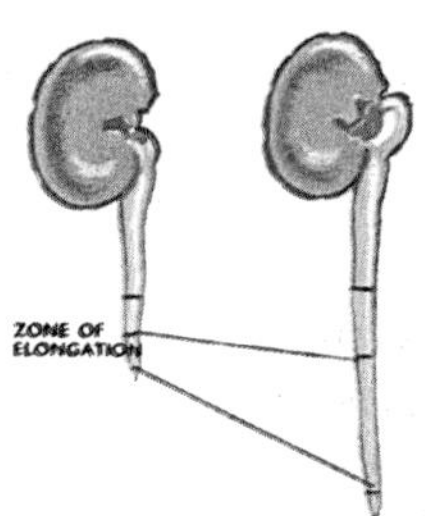

By marking a root or a shoot with equally spaced lines you can find out the region in which most of the growth takes place.

Roots grow in a similar way to stems. The tip of the root continuously produces new cells. As the base of the root is fixed, the tip is driven down into the soil. The tip is protected by a layer of cells called the *root cap*.

Many climbing plants have tendrils which grow rapidly when touched and coil tightly around a support.

Many flowers close up at night or when the weather is cold. The daisy closes its flowers when the light begins to fade in the evening. These movements are called nastic movements.

The Mechanism of Growth

A plant grows by producing new cells at the tips of its stems and roots. Most of the actual growth, however, takes place a little way behind the tip. However, growth is controlled by the tip which produces a substance called an auxin. This causes the cells to grow longer.

When a plant receives light from one side only, the light causes the auxin to stimulate growth on the shady side, so the plant bends towards the light. Plants also respond to gravity, so that stems grow up and roots down. Other movements include the bending of roots towards water, and the curling of tendrils around sticks. The movements are controlled by changes in water pressure in the cells. The leaves and flowers of some plants (for example, the nasturtium) bend during the day to keep facing the sun. They get the maximum amount of light in this way.

Vegetative Reproduction

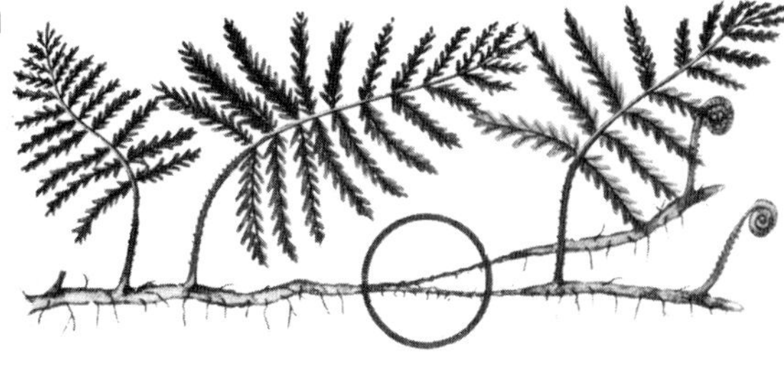

The more advanced plants generally have special reproductive organs and their reproductive processes involve the joining together of two cells. For example, among the flowering plants, pollen cells have to join with egg cells in the flowers before seeds can be formed (page 28). This method is known as sexual reproduction, but the process is not exactly the same in all plants.

In addition to the sexual process, many plants have a second method of reproduction. It is known as vegetative reproduction. Parts of the plant, usually the stem, become separated and grow into new plants.

The strawberry provides one of the best examples of vegetative reproduction. The plant sends out slender stems, called runners, that grow over the surface of the ground and then develop roots at their tips. Leaves develop there as well and the runners decay, leaving a number of new plants around the parent. Many plants spread in this way, but with underground stems instead of runners. These underground stems are called rhizomes. Many rhizomes store food as well. Other common methods of vegetative reproduction are pictured on this page.

Vegetative reproduction is an efficient way of spreading a plant over a small area, as we can see by the way in which couch grass or mint soon forms a large clump. But it does not carry the plants into new regions. A more important point is that the new plants are all the same, because they are all really just pieces of the parent plants. There is no variation of the sort we get when cells join together in sexual reproduction. Without variation there can be no improvement.

Bracken is a serious weed when it gets on to farm land because its underground rhizomes creep through the soil and soon cover a very large area.

Several plants, such as bryophyllum (near right) and coral root (far right), produce detachable buds or plantlets which grow into new plants.

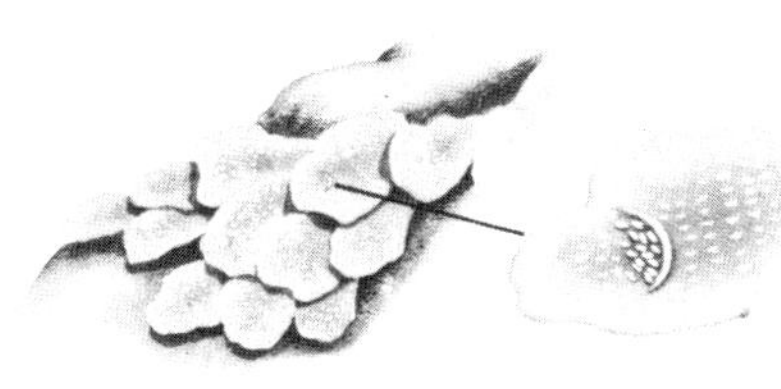

Some liverworts (page 21) produce detachable buds called gemmae. They are formed in little cups and they are washed away by the rain.

Strawberry runners grow very rapidly and soon cover the ground with young plants.

Animal Classification

The animal kingdom can be divided quite conveniently into two large 'parcels'—animals with backbones, and animals without backbones. These groups are properly called vertebrates and invertebrates, but they are very unequal groups. The vertebrates belong to just one of about 20 major groups into which the animal kingdom is divided.

These major groups are called *phyla* (singular *phylum*), and most of them are further divided into *classes*. The main phyla and classes are shown on this 'family tree' of animals. Some of the phyla have clearly descended from a common ancestor. Others are not clearly related to other groups.

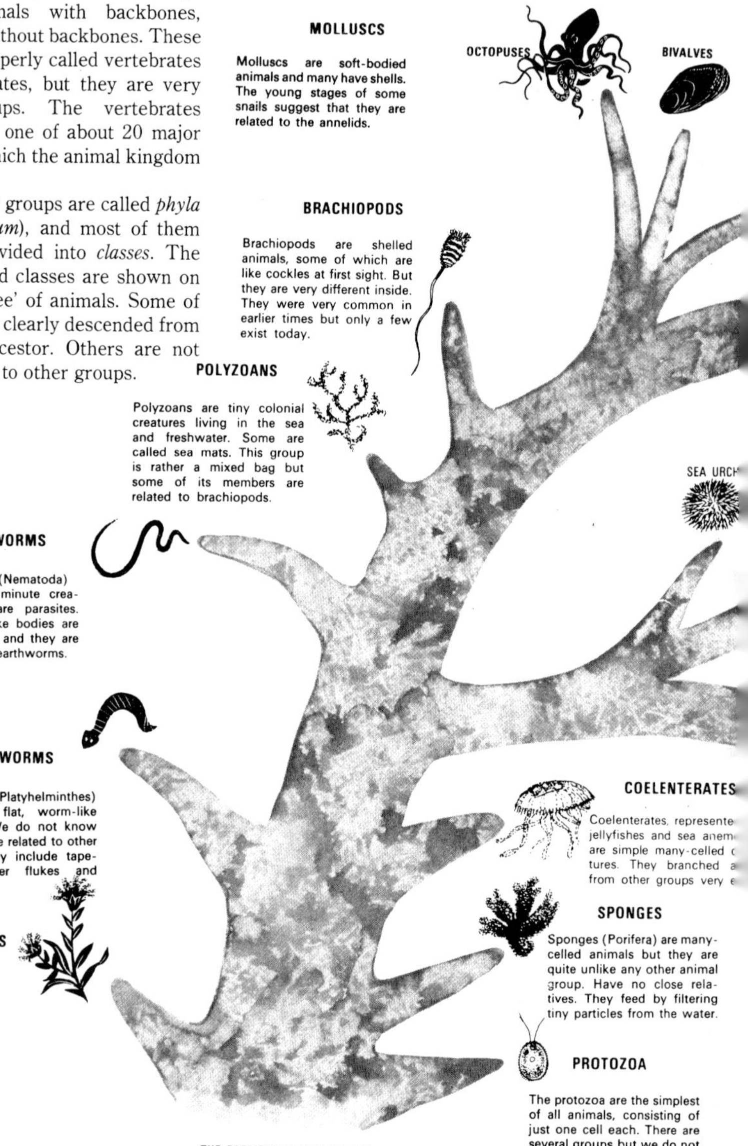

MOLLUSCS

Molluscs are soft-bodied animals and many have shells. The young stages of some snails suggest that they are related to the annelids.

BRACHIOPODS

Brachiopods are shelled animals, some of which are like cockles at first sight. But they are very different inside. They were very common in earlier times but only a few exist today.

POLYZOANS

Polyzoans are tiny colonial creatures living in the sea and freshwater. Some are called sea mats. This group is rather a mixed bag but some of its members are related to brachiopods.

ROUNDWORMS

Roundworms (Nematoda) are nearly all minute creatures. Many are parasites. Their needle-like bodies are not segmented and they are not related to earthworms.

FLATWORMS

Flatworms (Platyhelminthes) are rather flat, worm-like creatures. We do not know how they are related to other groups. They include tapeworms, liver flukes and planarians.

COELENTERATES

Coelenterates, represente jellyfishes and sea anem are simple many-celled c tures. They branched a from other groups very e

SPONGES

Sponges (Porifera) are many-celled animals but they are quite unlike any other animal group. Have no close relatives. They feed by filtering tiny particles from the water.

PROTOZOA

The protozoa are the simplest of all animals, consisting of just one cell each. There are several groups but we do not know how they are related.

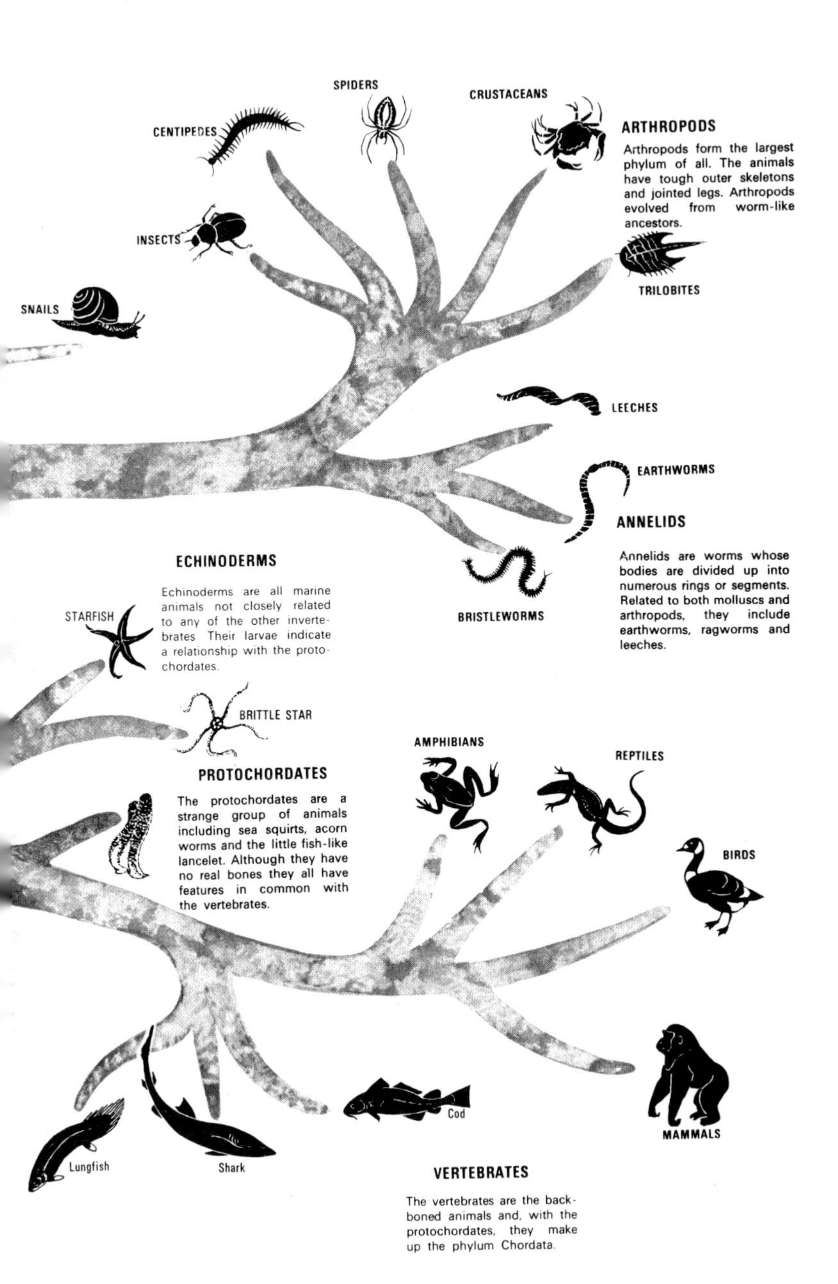

ARTHROPODS

Arthropods form the largest phylum of all. The animals have tough outer skeletons and jointed legs. Arthropods evolved from worm-like ancestors.

ANNELIDS

Annelids are worms whose bodies are divided up into numerous rings or segments. Related to both molluscs and arthropods, they include earthworms, ragworms and leeches.

ECHINODERMS

Echinoderms are all marine animals not closely related to any of the other invertebrates Their larvae indicate a relationship with the protochordates.

PROTOCHORDATES

The protochordates are a strange group of animals including sea squirts, acorn worms and the little fish-like lancelet. Although they have no real bones they all have features in common with the vertebrates.

VERTEBRATES

The vertebrates are the backboned animals and, with the protochordates, they make up the phylum Chordata.

The Microscopic World

Protozoa means 'first animals', and one kind, *Amoeba*, is popularly regarded as the most primitive form of life.

Amoeba's shape is always changing. Finger-like projections are continually being pushed out from the body and withdrawn and, as they move forward, the animal's body moves too. The whole animal therefore flows along. It can change direction merely by putting out a 'finger' or pseudopodium in another direction.

The animal also feeds by putting out these 'fingers'. They flow around small particles of food and completely engulf them. The particles are gradually digested and indigestible material is left behind as the animal flows forward.

Amoeba finds its food by a chemical sense and the animal reacts by moving towards food materials and away from harmful materials such as acids. The animal is also sensitive to light and it moves away from bright light. This sort of reaction ensures that the animal remains in suitable surroundings.

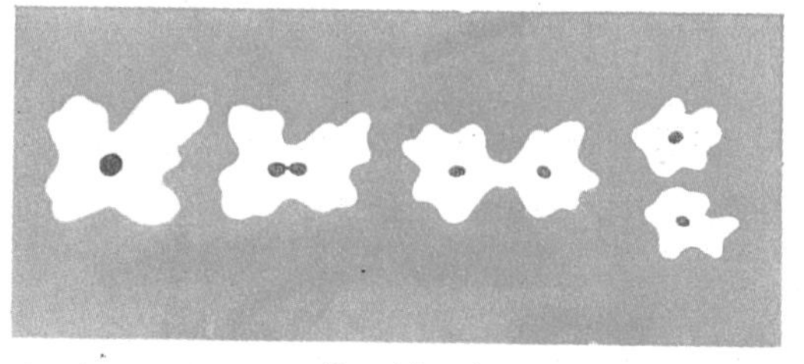

Amoeba merely splits into two halves when it gets too big. Each half then begins life as a new animal.

Euglena, one of the flagellates which are on the border between plants and animals.

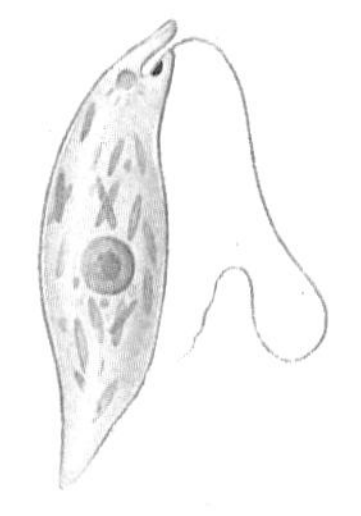

Breathing is no problem for the *Amoeba* because it is so small. Plenty of oxygen seeps in through its thin skin. The thin skin has, however, one disadvantage: the animal absorbs water from its surroundings.

The flagellates have one or more whiplike hairs called *flagella*, for swimming. Some flagellates, such as *Chlamydomonas*, are clearly plants. Some are equally clearly animals. But others, like *Euglena*, are on the border line between plants and animals.

Many protozoans live in the plankton of the sea. The foraminaferans have chalky shells and the radiolarians have glassy shells. Other protozoans live inside the bodies of other animals and cause diseases. Malaria is a serious disease caused by protozoans which are carried by mosquitoes.

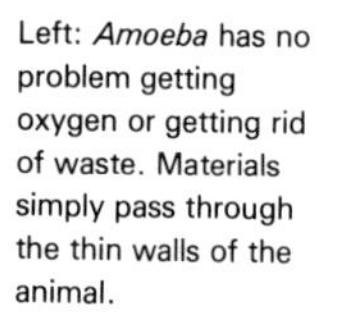

Left: *Amoeba* has no problem getting oxygen or getting rid of waste. Materials simply pass through the thin walls of the animal.

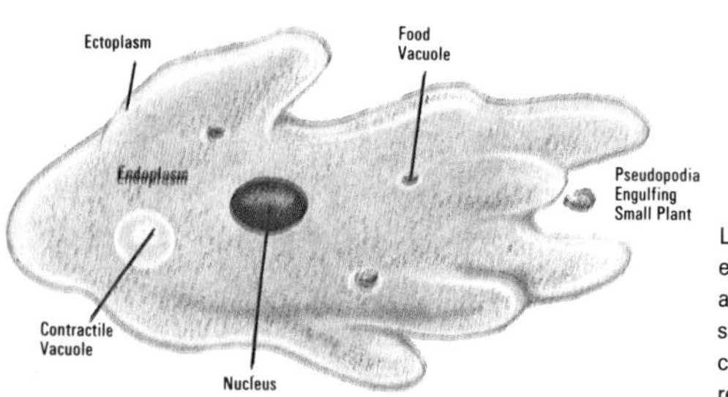

Left: A greatly enlarged *Amoeba* about to engulf a small plant. The contractile vacuole removes excess water.

Below: *Paramoecium,* another very common pond-dwelling protozoan.

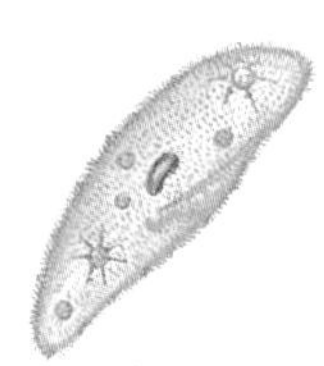

The Reef Builders

Part of a mature reef, with several different kinds of corals growing on it.

Coral reefs occur in the warmer seas of the world. The Great Barrier Reef off the coast of Queensland, Australia, is hundreds of kilometres long. It is hard to believe that such a massive structure has been built from the skeletons of tiny animals.

Corals are coelenterate animals closely related to the sea anemones. The main difference between the two is that the corals form limestone skeletons around themselves. Some corals are solitary, but the majority live in colonies.

The body of a coral animal is like a tiny sea anemone—a hollow tube with a number of tentacles surrounding the mouth. It is called a *polyp*. Some polyps reach a few centimetres across, but the majority are about the size of a grain of rice. The skeleton forms around the base of the polyp in the shape of a cup.

The coral polyps are carnivorous despite their small size. They feed on small crustaceans and other animals, which are trapped by the tentacles. These tentacles are armed with stinging cells, like those of the jellyfishes and sea anemones. They paralyse small animals which are carried by the tentacles to the mouth.

Corals and sea anemones reproduce by scattering eggs and sperms into the water. When the eggs are fertilized, they settle down to grow into new polyps. After a while the colonial forms begin to develop buds or branches. These grow out from the original polyp just above the rim of the skeleton. The latter then grows up around the branches as well. The older polyps die, but their skeletons remain to support the rest of the colony.

Solitary corals are found in most of the seas of the world, but the colonial species are found only in warm waters and the reef-formers are confined to areas in which the sea temperature never falls below 21°C, and where the water is clear.

There are three main kinds of coral reef. *Fringing reefs* grow close to the shore. *Barrier reefs* grow further away from the shore, and are separated from it by lagoons. *Atolls* are little coral islands, usually horseshoe-shaped, with a lagoon in the centre.

Whatever the type of reef, living coral forms only an outer coat, the rest is compacted skeletons.

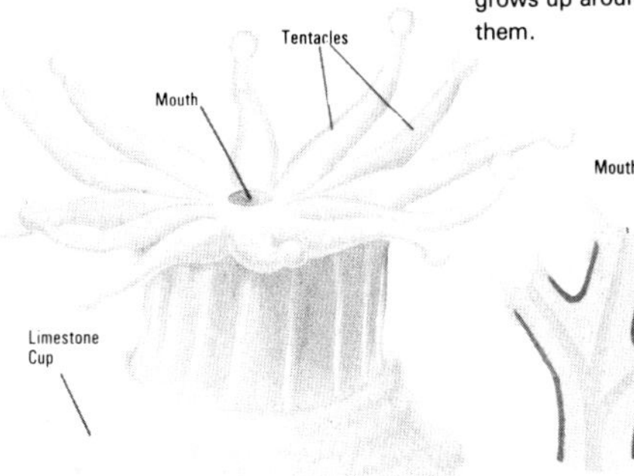

The coral starts off as a single polyp and forms a single cup around itself. New polyps branch off later and the skeleton grows up around them.

Worms and Leeches

We use the name worm for a whole range of long and slender animals. They belong to many different groups, but the most important are the flatworms, the round worms, and the segmented worms or annelids.

The Earthworm

The segmented worms get their name because their bodies are divided into a number of rings or segments. Huge numbers of earthworms live in the soil and their tunnels help to drain and aerate the soil, providing good conditions for plant roots.

The adult earthworm has a thickened region called the 'saddle' on the front half of the body. Each segment, apart from the first and last, has four pairs of tiny bristles on the lower surface, which help the worm to move.

Earthworms feed on decaying matter. They will often pull dead leaves into the soil, but they get much of their food simply by swallowing the soil as they tunnel. The undigested soil particles are passed out as *worm casts* on or near the surface. The worm thus brings mineral-rich soil to the surface.

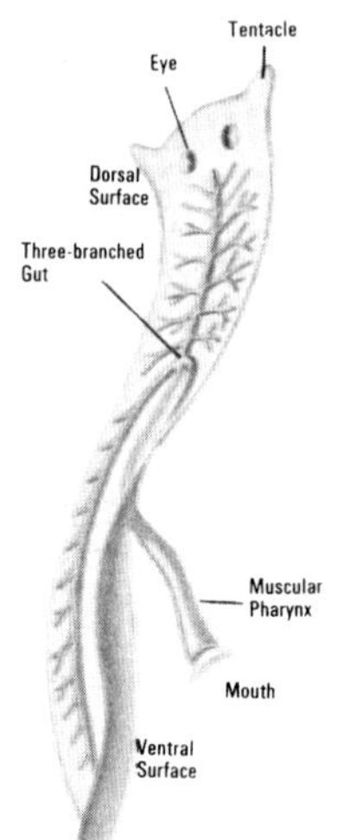

Planarian worms glide about in water. They have simple eyes on the head, but the mouth is near the middle of the body and can be pushed out on the end of a tube.

Below: Ragworms

Flatworms and Roundworms

The flatworms are rather primitive creatures. They include the little planarian worms that glide about the ponds. The flatworms also include the flukes and tapeworms, parasites that live inside the bodies of other animals.

Roundworms, or nematodes, are slightly more complex creatures. Many of them live as parasites inside other animals or plants. Those living in plants often cause serious diseases. Most of the roundworms, however, live freely in the soil or in water.

Bristle Worms

These are marine annelids, many of which have long bristles. Ragworms have well-developed heads with eyes and jaws. Other bristle worms live in the sand or mud, or else they form tubes of sand or lime. The lugworm burrows in the mud.

Leeches

Leeches have a large sucker at each end. Some are blood suckers; others eat small aquatic animals.

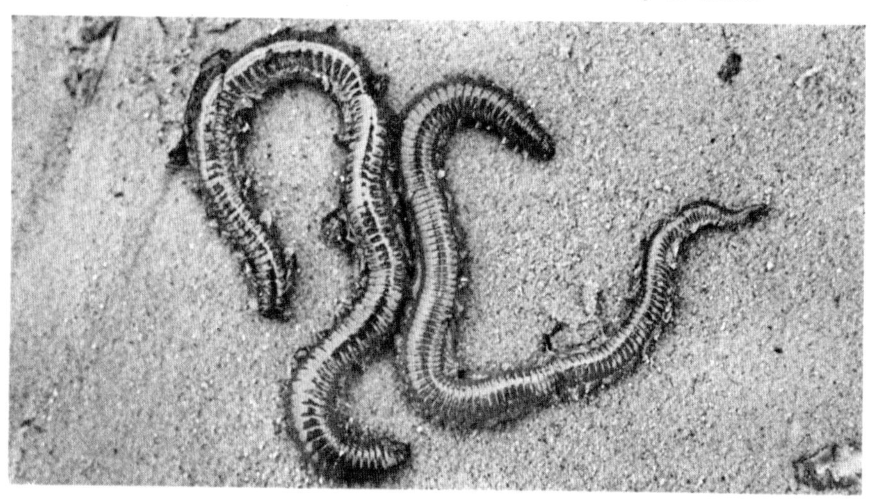

Crabs and their Relatives

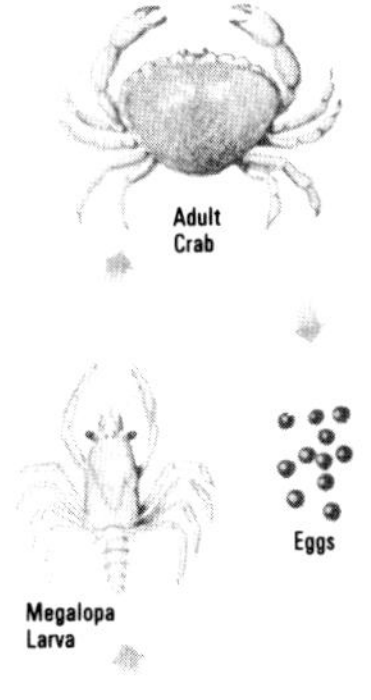

Crabs have a complex life cycle, often involving several stages that look nothing like the adult crab. They float freely in the surface waters.

Crabs, lobsters, shrimps, woodlice, and barnacles belong to the large group of arthropods known as the *crustaceans*. This name refers to the hard, chalky skin.

Except for the woodlice, the crustaceans live in water. This fact distinguishes them from most other arthropods, but there are other differences. For example, the crustaceans have two pairs of feelers or antennae. Many have a hard shell, or carapace, covering all or part of the body. There are usually many pairs of limbs, but they are of several different types, even in one animal.

The crustaceans range from minute floating creatures to lobsters weighing 17 kilograms and crabs with legs two metres long. Many live in fresh water, but the majority live in the sea. They make up a large part of the animal plankton—the drifting animals in the surface layers. Some crustaceans live permanently in the plankton. These include the shrimp-like krill which forms the bulk of the diet of the large whalebone whales. Another important planktonic crustacean, *Calanus*, is the main food of herring and other shoaling fishes. Many other crustaceans live in the plankton only during the early part of their lives. Crabs and lobsters have floating larvae.

Filter Feeders

The fairy shrimps have many pairs of broad, hinged limbs which filter food particles from the water while they 'row' the animal along. The limbs also act as gills and absorb oxygen from the water. Related to the fairy shrimps are the water fleas. They swim by 'rowing' with their large second pair of antennae. This produces a jerky upward movement, from which the animals get their name.

The common barnacle is closed when the tide goes out. When the water returns, the animal opens and 'combs' food from the water with its delicate limbs.

Crabs, Shrimps, and Lobsters

These animals are often called *decapods* because they have ten walking legs. At least one pair is modified to form pincers. Crabs are relatively short, broad animals, in which the tail region is folded tightly under the body. The shrimps, prawns, and lobsters have longer tails with several pairs of legs called *swimmerets*.

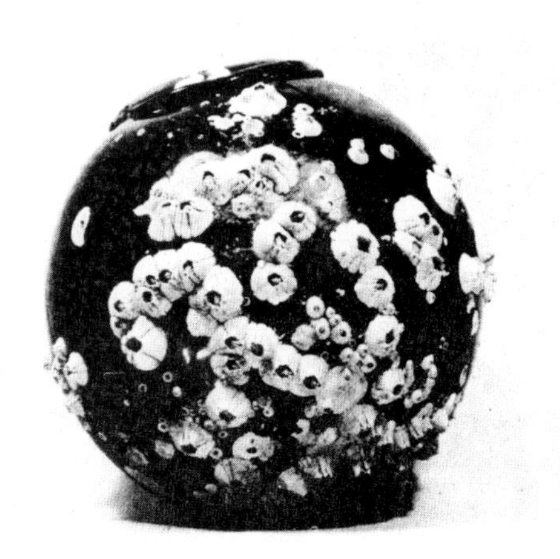

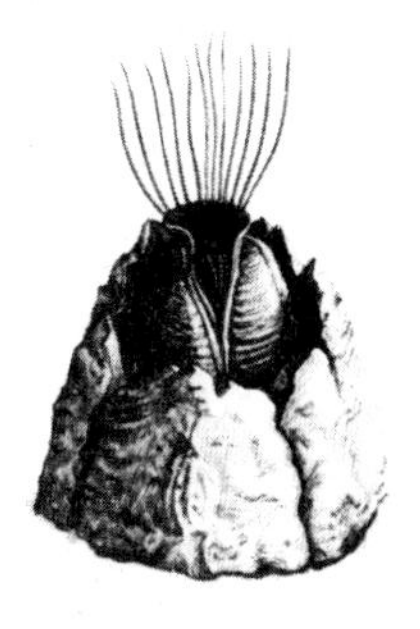

Left: These barnacles coating a glass float look rather lifeless out of the water, but when submerged by the tide the inner plates open up and out come the barnacle's limbs (above) to comb food from the water.

The Insect World

The insect world has more than a million species and untold millions of individuals. Insects can be distinguished from other arthropods because they have three distinct regions to the body. These are the head, the thorax (in the middle), and the abdomen. The thorax bears three pairs of legs and it usually carries two pairs of wings. Insects also have a single pair of feelers, or antennae, on the head.

Insects live almost everywhere on the earth—on the land and in the water, in the deserts and around the poles. Few live in the sea.

One of the features that prevents the insects from getting too large is their method of breathing. Tiny tubes carry air from the surface to all parts of the body, but the system works only over short distances and so insects cannot grow very large.

Insect Groups

There are about 30 major groups of insects, separated mainly by differences in the wings and the methods of feeding. Four groups contain rather primitive insects without wings. These include the little silverfish. The other groups are basically winged insects. They all had wings at some time during their history, although some have lost them again, like the parasites that live among fur and feathers.

Below: The compound eyes of a fly. The numerous tiny lenses form a honeycomb-like mosaic. Each lens sends its own signal to the brain so that the insect sees an image as a pattern of dots. The image is not very clear and insect eyes are best suited to picking out moving objects.

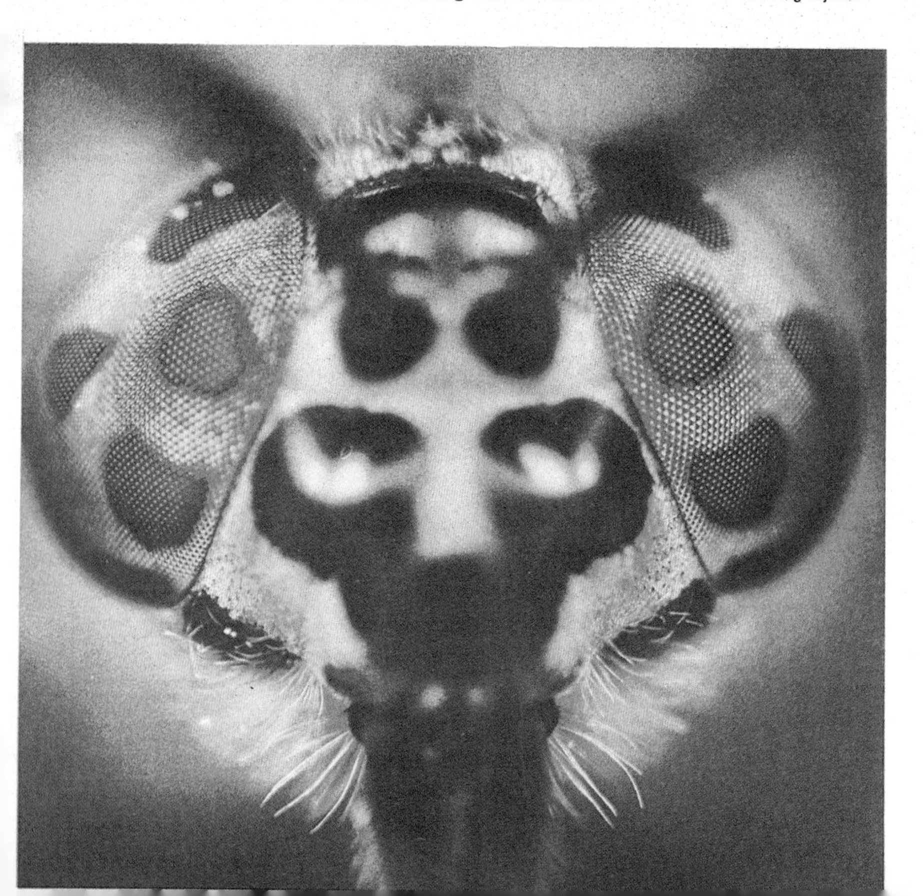

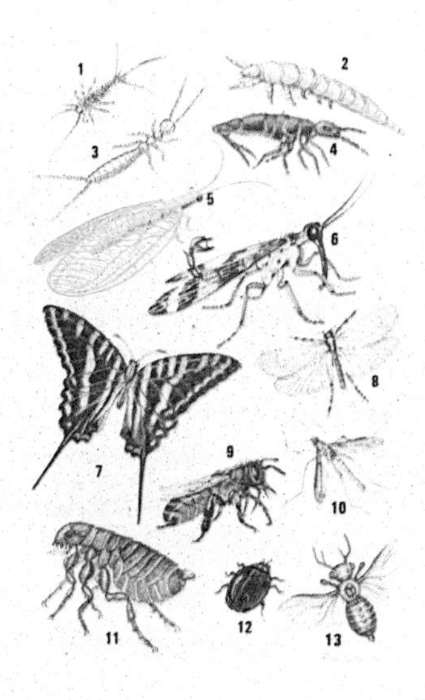

Some important insect groups. 1, Order Thysanura includes the bristletails and silverfish. 2 and 3, Orders Protura and Diplura. 4, Order Collembola, the springtails. 5, Order Neuroptera, lace-wing flies. 6, Order Mecoptera, scorpion flies. 7, Order Lepidoptera, butterflies and moths. 8, Order Trichoptera, caddis-flies. 9, Order Hymenoptera, bees, wasps, and ants. 10, Order Diptera, true flies. 11, Order Siphonaptera, fleas. 12, Order Coleoptera, beetles. 13, Order Strepsiptera includes parasites of bees.

Insect Life Histories

A young grasshopper looks very much like an adult grasshopper, except that it is much smaller and has no wings. It is called a nymph. Because the skin is rigid, the insect cannot grow except by shedding the skin, or moulting. The skin splits and the insect then wriggles out. A new and larger skin has already grown beneath the old one. It is soft and the insect puffs itself up with air before the new skin has hardened.

After the first moult the grasshopper nymph begins to develop its wings, but they do not become fully formed until after the fourth moult. The insect is then a fully grown adult grasshopper and it does not moult any more. This sort of life history, in which the young insect gradually takes on the appearance of the adult, is also shown by dragonflies, earwigs, and bugs. It is called partial or incomplete metamorphosis.

Butterflies, bees, houseflies, and beetles have a different kind of life history. The young stages of these insects are called grubs or caterpillars, and they look nothing like the adult insects. They often eat quite different food from the adults. The larvae change their skins, but there is no sign of wing flaps. Then the larvae turn into pupae. The pupae do not feed and they do not usually move much, but great changes are going on inside. The young insect changes into the adult inside the pupa. This abrupt change is called a complete metamorphosis. When the adult comes out of the pupa, its wings are crumpled and its body is soft, but it is soon ready to fly.

Insect Diets

Many insects feed on plants. Well-known examples include caterpillars (young butterflies and moths), green-fly, and many beetles. Other insects are carnivorous. Dragonflies, bush crickets, and ladybird beetles eat other insects. Fleas, bedbugs, and mosquitoes suck the blood of much larger animals. Then there are the omnivorous insects, like the cockroach, which eat more or less anything they can find.

Right: A lesser stag beetle shows the major features of insect anatomy—three body sections, three pairs of legs, and one pair of feelers.

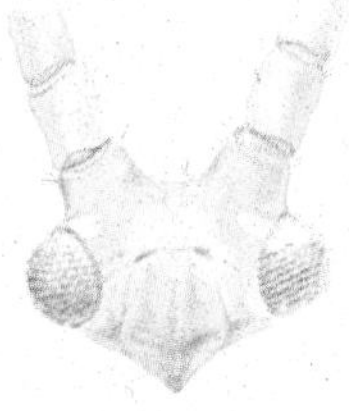

Above: The head of an aphid showing the compound eyes.

Insect Mouth-parts

Grasshoppers, beetles, and wasps have biting jaws for eating solid food. Other 'legs' around the mouth help hold food and push it into the mouth.

Butterflies and moths have a long 'tongue', or proboscis, which unrolls to suck nectar. Bugs and fleas have sharp, needle-tongues which they sink

The powder post beetle is one of several species whose grubs do much damage by tunnelling into wood.

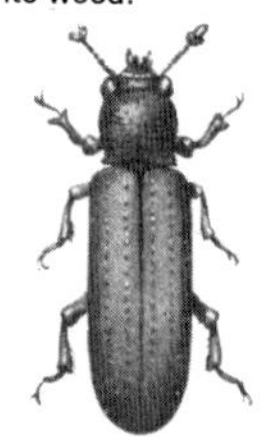

into plants and animals to drink sap or blood.

Many flies have sharp, piercing mouth-parts for sucking blood. The mosquito is well known, and others include the tsetse fly, stable fly, and horse fly. Like fleas and bugs, many of them carry diseases from one animal to another. The housefly and the bluebottle, however, have different kinds of mouths. They form spongy pads and the insects use them to suck liquid food. Solid food can be dealt with by pouring digestive juices on to the food and then sucking up the resulting solution. The digestive juices come from the food canal and often contain parts of the fly's previous meal. The flies can contaminate our food with germs picked up somewhere else.

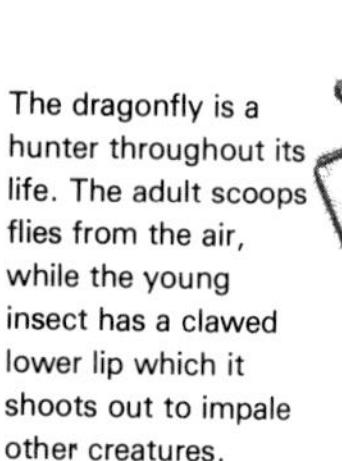

The dragonfly is a hunter throughout its life. The adult scoops flies from the air, while the young insect has a clawed lower lip which it shoots out to impale other creatures.

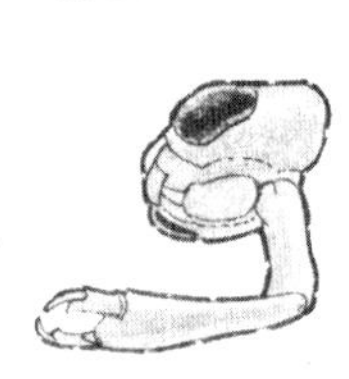

Below: The housefly mops up liquid food with its sponge-like tongue.

Below: The mosquito female is a blood sucker. She sinks her needle-like mouth-parts into other animals.

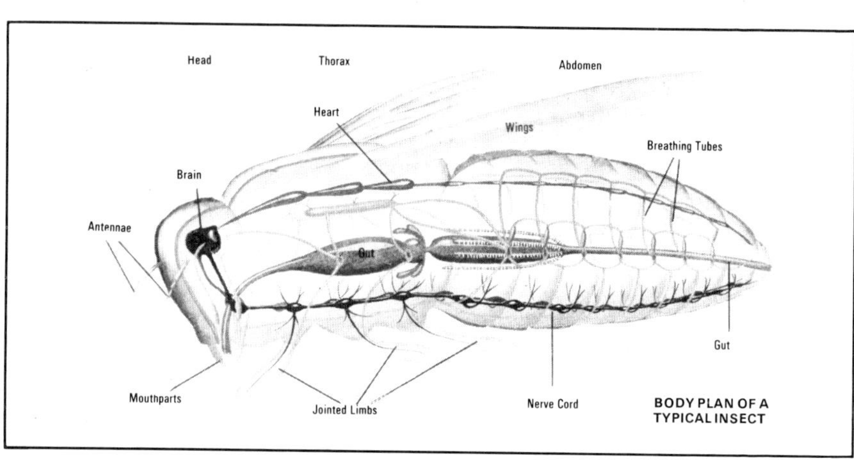

BODY PLAN OF A TYPICAL INSECT

Insect Pests

Clothes moths eat our clothes and carpets, furniture beetles (woodworms) tunnel through our furniture, grain weevils destroy stored grain, caterpillars eat our cabbages, mosquitoes spread malaria, and houseflies contaminate our food. These are just some of the insect pests which man has to tolerate.

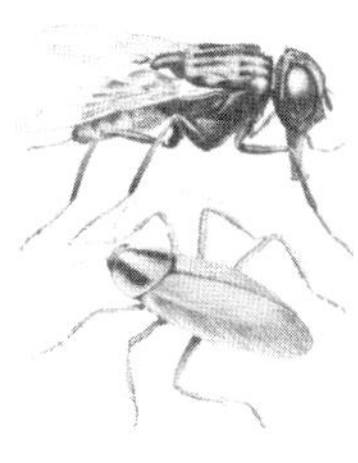

Above: Two domestic pests, the housefly and the cockroach.
Below: The Colorado beetle from America, a widespread pest of potatoes.

Domestic Pests

One of the commonest household insects is the housefly. It breeds in all sorts of rubbish, and it finds suitable breeding sites in the dustbins and rubbish tips of human settlements. The adults lay their eggs on the rubbish and they also feed there. Then they contaminate our food with germs picked up on the rubbish dump. The use of insecticides, together with better methods of refuse disposal, has reduced the housefly population in recent years.

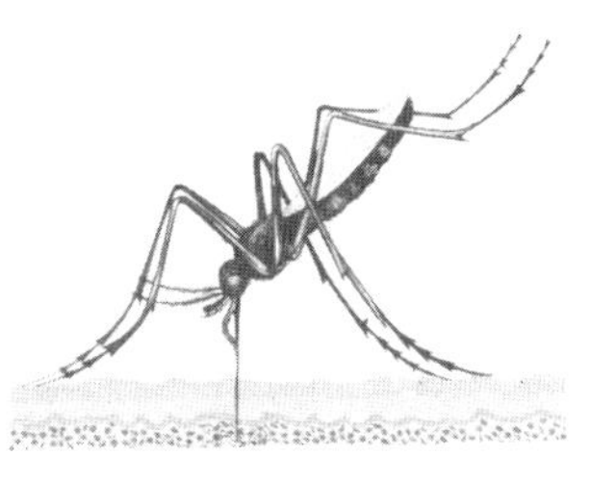

Left: A feeding mosquito plunges its 'beak' into the skin of an animal. Mosquitoes are serious pests because they carry the germs of malaria, yellow fever, and other diseases.

The caterpillars of clothes moths feed on woollen articles. Many clothes moths have been found in the nests of birds and rodents, where they feed on scraps of feather and fur. It is quite natural, therefore, that they should invade the homes of men in search of similar materials. Carpet beetles have had much the same sort of history.

The furniture beetle or woodworm and the much larger death-watch beetle make their homes in dead wood. Their larvae tunnel in dead tree-trunks, but they are also quite happy to feed on our furniture and floorboards.

Carriers of Disease

Mosquitoes, tsetse flies, fleas, and lice are responsible for carrying disease-causing germs. Most of them are blood-sucking insects and if they take in germs when they feed on one person they are quite likely to inject some into a later victim. Malaria is carried only by certain kinds of mosquitoes, while sleeping sickness is normally carried only by tsetse flies.

Crop Pests

A great deal of the world's food is lost every year because insects eat it or damage it while it is growing. The Colorado beetle came originally from the western parts of North America, where it fed on wild plants and did no harm to anyone. Around the middle of the nineteenth century, people started to grow potatoes in the region, and from then on the Colorado beetle has been a pest. The potato plant is related to the wild plants on which the beetle fed. Consequently the Colorado beetle soon transferred its attention to the potatoes. Aphids—greenfly and blackfly—are probably the most serious insect pests of crops. There are many different kinds and they affect a whole range of plants. They weaken plants by sucking sap, but they do even more damage by carrying virus diseases. The diseases weaken the plants even further and may kill them.

The large white butterfly caterpillars feed on cabbages and do a great deal of harm to crops throughout Europe.

Butterflies and Moths

Butterflies and moths belong to a large group of insects called the Lepidoptera. This name means 'scale wings' because the wing is covered with tiny overlapping scales. These scales give the wings their patterns. Some scales in male butterflies also give off scent which attracts the female.

The Lepidoptera is one of the largest groups of insects and contains something like 100,000 different species, most being moths. There is an enormous size range within the moths, from insects such as the giant owlet moth of South America (30 cm from wing tip to wing tip) to minute moths no more than 3 mm across whose caterpillars live between the upper and lower surfaces of leaves.

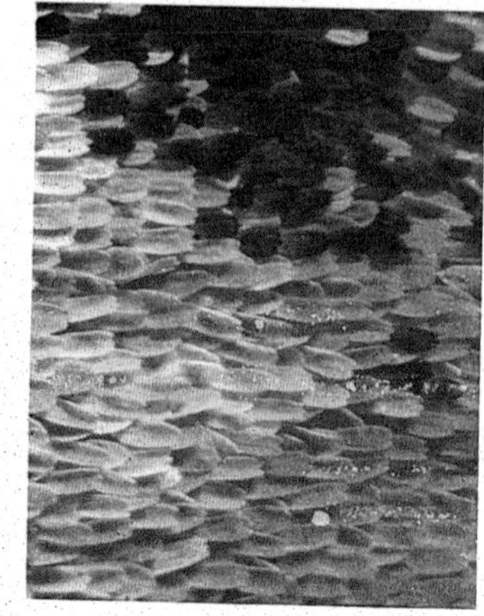

Right: The colours of a butterfly's wings are produced by thousands of tiny overlapping scales which come off if the wings are rubbed.

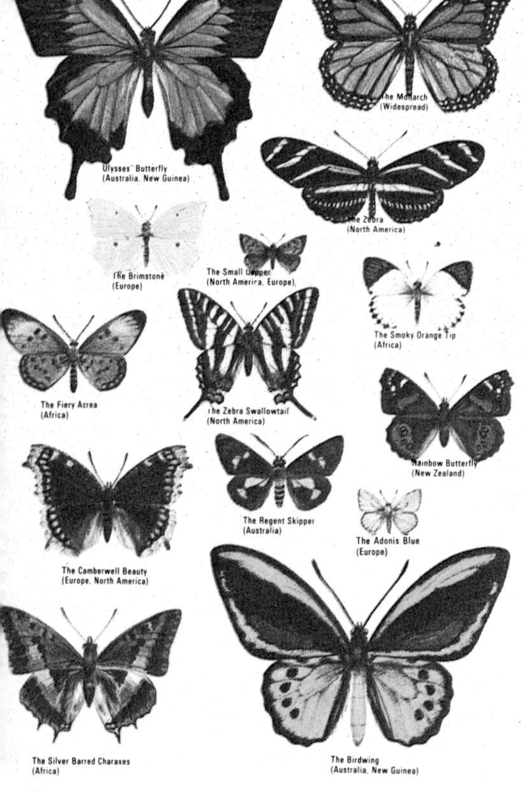

Below: Butterflies vary a great deal in shape and size. These species, belonging to several families, are all half life size.

There is no simple distinction between butterflies and moths. It is often said that butterflies are brightly coloured and fly by day, whereas moths have dull colours and fly at night. This might be true for the majority of species, but certainly not for all of them.

The best way to distinguish a butterfly from a moth is to look at its feelers or antennae. All butterflies have clubbed antennae. Moths have many different kinds of antennae, but rarely have clubbed tips. Those moths that do have clubbed antennae can be distinguished from butterflies by looking at the wings. The 'shoulder' of the hind wing carries a bristle, known as the *frenulum*, which fits into a little hook on the underside of the front wing. This arrangement holds the two wings together in flight. Butterflies do not have a frenulum. There are many moths without a frenulum, but these never have clubbed antennae.

Reproduction

The butterflies and moths belong to that division of insects which undergo a complete change, or *metamorphosis*, during that time. The eggs are usually laid on the insect's food plant and they are often beautifully patterned. All caterpillars have biting jaws and, with the exception of a few 'clothes moths' which feed on woollen material, they are vegetarians. Some feed inside the stems, roots, and fruits of plants, but most feed on the leaves.

When ready to pupate, many caterpillars, especially those of moths, spin silken cocoons around themselves and attach them to the branches of the food plant. Others hollow out snug chambers in the soil, while many butterfly larvae simply attach themselves to a convenient support by a pad of silk. Inside the pupa, the caterpillar's body is broken down and rebuilt into the body of the adult insect. This may take as little as 10 days, but it is often nine months before the adult insect emerges. When the adult first crawls out through a split in the pupal skin it is very soft and its wings are tiny crumpled objects. Its first action is to climb up some support and hang itself up to dry. It pumps blood into the veins of the wings and they gradually expand to their full size. After a few hours they are hard enough to bear the insect away on its first flight.

Sight and scent both play a part in the courtship of butterflies and moths. Among butterflies it is the male who gives out the scent, but among moths it is usually the female. Some male moths have extremely sensitive 'noses' on their antennae and can smell a female from more than a kilometre away.

Feeding

A few very primitive moths have biting jaws, which they use for feeding on pollen, but the rest of the moths and the butterflies have lost their jaws. Their feeding apparatus, if they have any, consists of a slender tube which they use to suck nectar and other sweet substances.

The majority of butterflies and moths are quite harmless in the garden, and many are very useful because they pollinate the flowers and help them to set seed. There are some species, however, which are garden pests. The 'cabbage whites', for example, are found in most parts of the world and their caterpillars destroy valuable greenstuff. Bufftip moth and lackey moth caterpillars damage fruit trees by stripping their leaves. The gypsy moth, a European species, created havoc in forests when it reached North America.

Bright colours will not always distinguish butterflies from moths—the tiger moth at the bottom is just as bright as the fritillary butterfly. The antennae are a better guide: almost all butterflies have distinctly clubbed antennae.

Ants and Termites

There are more than 10,000 different kinds of ants. All ants live in colonies and, like the honey bee, they have three main castes. The colony is headed by a mated female called the queen. Male ants are produced at certain times of the year, but most of the ants in a colony are sterile, wingless females. These are the workers and they are the most familiar of the ants—they are the ones we see crawling over the ground and the vegetation in search of food. Many ant species also have an extra caste called a soldier. This is a large-headed ant whose job is to defend the colony with its large jaws. Some of the soldiers also use their jaws to crack open the hard seeds on which some ants feed. The ordinary workers help in the defence of the colony by using their stings and sharp jaws. Not all the species have stings, but some squirt formic acid at their attackers.

The typical features of ant life are shown by the black garden ant, *Lasius niger*, although this species does not have a soldier caste. Winged male and female ants are produced during the summer and, when the weather is just right, they emerge for their wedding flights. All the nests in an area will release their flying ants together, so we find great swarms of them at this time. The males die after the wedding flight, but the females or queens begin their duties as nest founders and mothers. They break off their wings and find suitable holes in which to hide until their eggs are ready to be laid.

Above: The wood ants' nest has many chambers. Eggs are kept in one part, larvae in another, and so on. The queen ant can be seen in her central royal chamber.

Above: Some ants can turn their hind ends forward and squirt acid at their attackers.

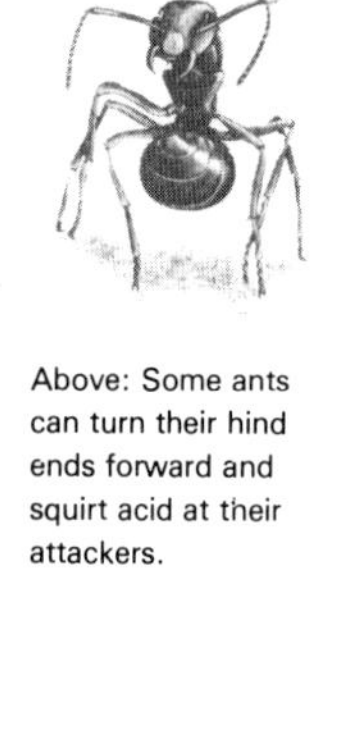

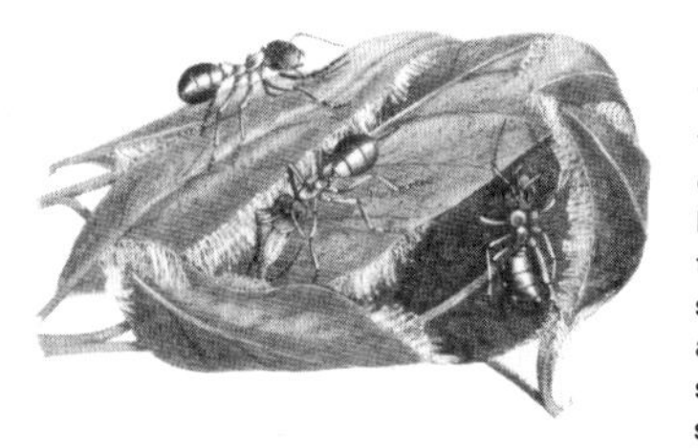

Left: Weaver ants make small nests by sewing or gluing leaves together. While some ants hold the leaves together other ants use the larvae rather like tubes of glue. They squeeze the larvae and make them secrete strands of sticky silk.

When the first eggs hatch the queen feeds the larvae on saliva. They soon turn into worker ants and set to work to build the nest. This consists of numerous chambers and passageways. The queen remains in one chamber and devotes the rest of her life to laying eggs. She may live for several years, although the workers do not live so long.

As well as building the nest, the workers collect food and look after the young ants. Plant and animal material is eaten, but many ants are particularly fond of honeydew. This is the sweet, sticky liquid given out by greenfly. The ants stroke the greenfly which makes them give out honeydew, and the ant then takes it back to the nest.

Some tropical species make no permanent nests at all. They are called army ants or driver ants. They settle down for short periods now and then to rear more young, but most of the time they are on the march. A marching colony may be several metres wide and hundreds of metres long and its members eat any animal in its path.

Many ants harvest seeds and herd aphids, but some actually grow their own crops. These farming ants live in South America and they feed entirely on fungi which they grow on beds of chewed leaves in their nests. The workers go out and bring back pieces of leaves, waving them above their heads as they go. For this reason they are often called parasol ants.

Slave Makers

Some species of ants have no workers of their own. They lay their eggs in the nests of other ants and rely on these ants to rear them. Other species have some workers but rely on 'slaves' to do most of their work. Workers of the slave-making ants raid the nests of other ants and bring back pupae. The ants from these pupae are the slaves.

Termites

Termites are often called white ants, but they are not related to the ants. Termites live in the warmer parts of the world and there are about 2,000 species. Two species live in southern Europe. They feed mainly on wood, and some species do great damage to trees and to buildings. Wood is difficult to digest, and the termites enlist the help of tiny protozoans which live in the termite's food canal and digest the wood for it. In return, the protozoans take some of the food for themselves.

Right: Parasol ants carrying leaf fragments back to the nest.

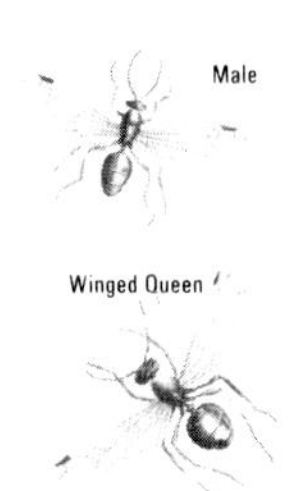

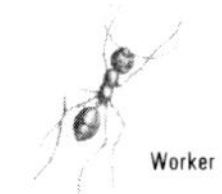

Above: The three castes of the wood ant.

Below: The queen termite is swollen with eggs. The king rests on her back, while they are protected by soldiers.

All the termites are social creatures and, unlike the ants, they have a king and a queen at the head of the colony. The workers and the soldiers are also of both sexes.

Termite colonies are started by both the king and the queen after they have had a short flight. The workers take over later and the royal couple retire to the royal chamber, where they may live for 50 years in some species. The most primitive termites make their nests in dead wood—usually in old tree trunks or logs. Other termites make their nests under the ground, and the soil dug out may be heaped into huge mounds and cemented with the insects' saliva. These mounds are called termitaria and they are especially common in Australia and South Africa. They are full of chambers and tunnels and have hard walls which protect the termites from hot weather.

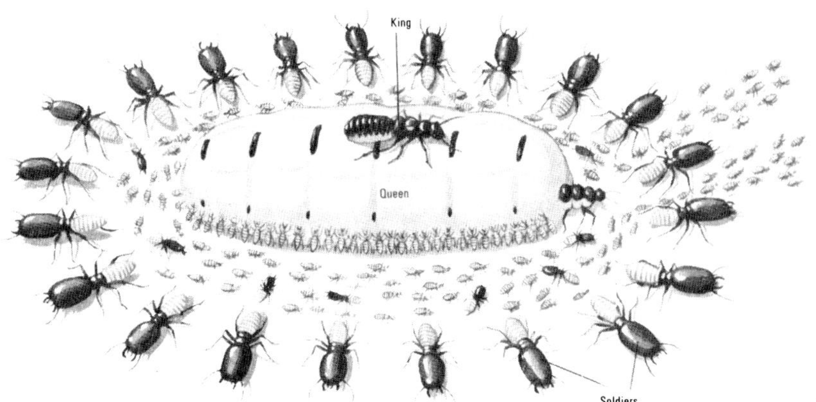

The Honey Bee

Most of the insects live alone and look after none but themselves. Some of the bees and wasps, however, and all of the ants and termites live in communities of many individuals. They are social insects and each individual works for the good of the whole community.

The honey bee came originally from South-East Asia, but it has been taken to many parts of the world to provide men with honey. Wild honey bees normally nest in hollow trees, but they are quite happy to make their homes in the bee keeper's hives.

Ruled by the Queen

The honey bee colony is headed by the *queen*. She does nothing but lay eggs. Most of the bees in the colony are *workers* which are sterile females. Although the workers do all the work, the smooth running of the colony is maintained by the queen. She produces an oily material called *queen substance* which the workers lick off and spread around among themselves. It maintains the 'community spirit'. A third type of bee is produced in the colony. The male or *drone* appears during the summer. He mates with a new queen when one appears.

Life in the Hive

The bees' nest consists of a number of 'combs' hanging vertically in the hive or tree. Each comb is composed of hundreds of little six-sided cells, all made of wax from the bees' bodies. The cells are used for storing pollen and honey and also for rearing the young. The queen lays an egg in each empty cell. When the eggs hatch the workers feed the grubs. For the first three days the grubs receive a substance called *brood food*. This is a protein-rich material made by the young workers. Then they are fed on pollen and nectar. About nine days after the eggs were laid the grubs turn into pupae, and the adult bees emerge about 10 days after that.

Bees collect pollen and nectar from flowers. Pollen clings to the bees' hairs and is brushed into baskets on the back legs.

Below: The direction of the bee's dance tells the other bees where to find nectar in relation to the sun's position.

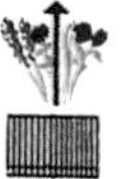

The young workers spend a few days doing 'housework' and feeding the grubs. Then they take over the duties of making and repairing the combs. Not until about three weeks after emerging from their cells do the young workers go out and search for food.

The bees visit flowers and collect both pollen and nectar. While doing this they pollinate the flowers. This pollinating activity is more important to us than the honey that the bees provide. The bees carry pollen back to the hive in the pollen baskets on their back legs. Nectar is given up to other bees in the hive who remove much of the water from it and convert it into honey.

Bumble Bees

Bumble bees do not live in such large colonies as honey bees. The colony lasts for only one year. New queens are produced in the summer and, after mating, they hibernate. They wake up and build small nests in the spring. The drones and workers all die in the autumn. The social wasps have a very similar life story, with only the queens surviving the winter. Wasps, however, feed their young on insects instead of pollen and nectar.

Bee Dances

The bees have ways of telling each other about good nectar sources that they have found. They come back to the hive and 'dance' excitedly on the combs. The type of dance tells the other bees where to find the nectar. They find their way by the sun and they even make allowances for the movement of the sun across the sky during the day.

Swarming

Every now and then the population in the hive gets rather too large for comfort and the workers prepare for a swarm. They build special queen cells at the edge of the comb. The queen lays an egg in each queen cell and the grubs are fed on brood food for the whole five or six days of their larval life. The extra protein that they receive ensures that they become fully developed females or queens. The first queen to emerge from her cell kills the other developing queens and then goes off on a marriage flight with one or more drones. She returns to the hive later to become queen. In the meantime her mother, the old queen, has left the hive with a swarm of workers to start a new colony somewhere else.

The queen honey bee normally lives for about three years and she lays over half a million eggs during this time. The workers live for only a few weeks during the summer, although those reared in the autumn will survive until the spring. In winter the colony is quite small and it survives on the food stored up during the previous summer. Numbers build up again in spring when food becomes plentiful, and there may be 60,000 bees during the summer.

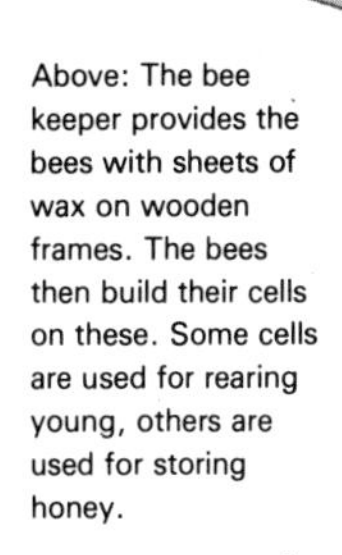

Above: The bee keeper provides the bees with sheets of wax on wooden frames. The bees then build their cells on these. Some cells are used for rearing young, others are used for storing honey.

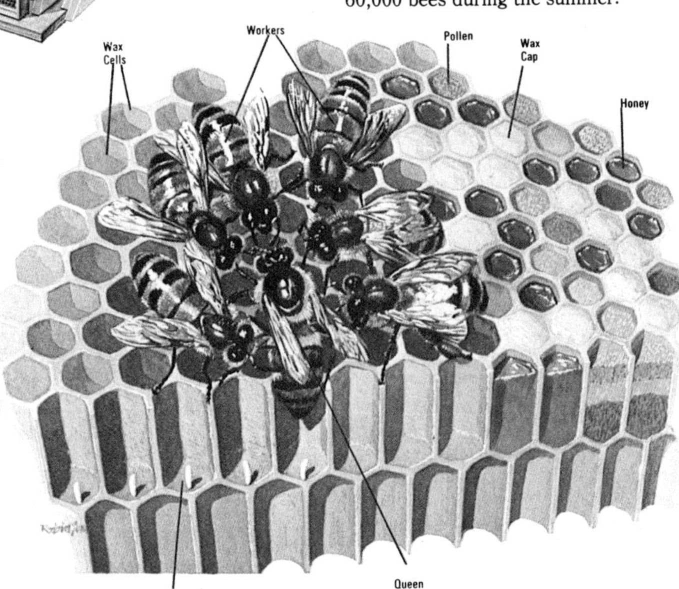

Right: The queen wanders over the combs and lays an egg in each empty cell she finds. She is always attended by workers, who lick her and touch her with their feelers to obtain queen substance.

Spiders and Scorpions

Spiders and scorpions, together with mites, ticks, and a few other animals, form the group of arthropods called *arachnids*. They normally have four pairs of walking legs.

Spiders

The spider's body is divided into two regions. The front part bears several pairs of eyes and it also carries the legs. Around the mouth there is a pair of poison fangs and a pair of small leg-like palps.

Spiders eat only freshly caught animals. Most eat insects, but some of the larger ones eat birds, lizards, and small mammals. The spider's venom kills or paralyses the prey, and digestive juices are then poured into it. The spider then proceeds to suck it dry.

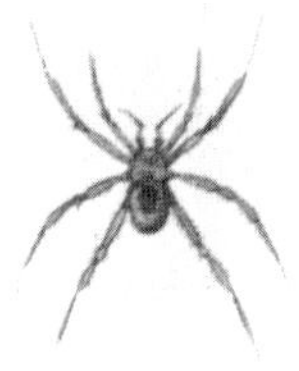

Above: A harvestman, distinguished from spiders by the absence of a 'waist'.

The Spider's Web

Some spiders chase their prey, or lie in wait, but many make traps or snares in the form of silken webs. There are many different kinds of webs, but the circular orb-webs spun by the garden spider and its relatives are the most familiar. First of all it makes a frame and a number of radii, like the spokes of a wheel. Then it moves round on the radii and leaves a spiral of dry silk. Next, starting from the outside and using the first spiral as 'scaffolding', it lays down a tighter spiral of sticky silk. This is what traps the insects.

All spiders produce silk, whether they make webs or not. It is used for wrapping up the eggs, and also as a life-line for a spider that falls from its home. It pays out a thread as it falls, and then simply climbs back up. Young spiders also use silk to help them to travel. They let a strand blow in the breeze until it whisks them up into the air.

The silk is produced in the abdomen. The silk glands open through tiny pores called spinnerets near the hind end. The silk is liquid to start with and it flows out like toothpaste from a tube, but it hardens as soon as it meets the air. Male spiders are usually smaller than the females and they have to take great care when they go courting. So that they are not eaten by the female, they signal by waving their palps or tugging her web.

Below: A dew-spangled web, death trap for unwary insects.

The Water Spider

Spiders are basically land animals. Several skate across the surface, but only one species lives under the surface. The water spider lives in a silken tent under the water which it fills with air brought down from the surface.

Harvestman

The long-legged harvestman is often confused with a spider. The most obvious difference is that the harvestman's body is not divided into two parts. It feeds on insects and other small animals. Unlike the spiders, it cannot make silk.

Scorpions

In these animals, the palps are powerful pincers and the abdomen is prolonged into a tail with a sting at the end. Scorpions eat insects and other small animals, which they catch with their pincers. The food is then torn up by the smaller pincers at the front of the head and the juices are sucked up. Most species are active at night and hide under rocks and stones by day.

Slugs, Snails and Bivalves

Slugs, snails, cockles, mussels, oysters, and octopuses are *molluscs*. They are all very soft-bodied animals without backbones. Most of them are protected by a hard shell.

Snails

Snails belong to the group of molluscs called *gasteropods*. They usually have a coiled shell into which they can retreat. Many also have a horny plate with which they can close the shell. Snails live in the sea, on land, and in fresh water.

The snail's body has three main regions. The most obvious is the flattened *foot* on which the animal glides about. It is propelled by muscular ripples. A gland near the front of the foot pours out slime which lubricates the snail's path and makes movement easier. The *head* bears tentacles. The eyes are at the tip of the larger tentacles in land snails, and at the base of the tentacles in other species. Most of the snail's internal organs are in the coiled *hump*, which remains inside the shell. The hump is covered by a thick 'clock' called the *mantle*, which forms the shell.

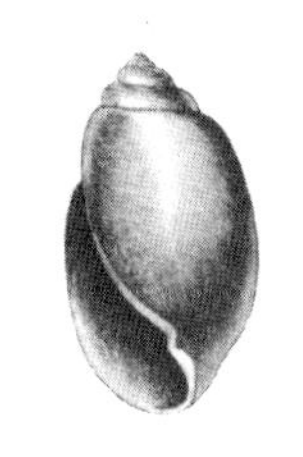

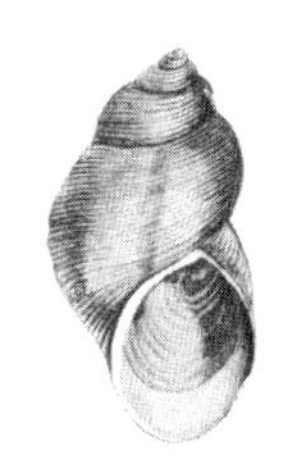

Above: Most snail shells are right-handed (as the lower shell) but a few are left-handed (like the upper shell).

Most aquatic snails breathe with gills. The gills lie in a cavity under the mantle. The mantle cavity of land snails has been converted into a lung. The snails pump air in and out of the cavity instead of water. Many water snails also breathe in this way: they surface every now and then to take in a lungful of fresh air.

Most of the land snails are vegetarians, feeding on dead plant material as a rule. Freshwater snails are also mainly plant-eaters, and so are many marine species. Other marine snails, however, are carnivorous. The whelk, for example, feeds on other molluscs. Snails have a peculiar 'tongue' called a *radula*, covered with hundreds of horny teeth. It is used like a file to rasp away the food. The teeth are always wearing out, but the radula keeps growing and producing new ones.

Most snails are hermaphrodite animals. This means that each snail has both male and female organs. They have to pair up, however, before they can reproduce. Some snails lay eggs, and others give birth to active young.

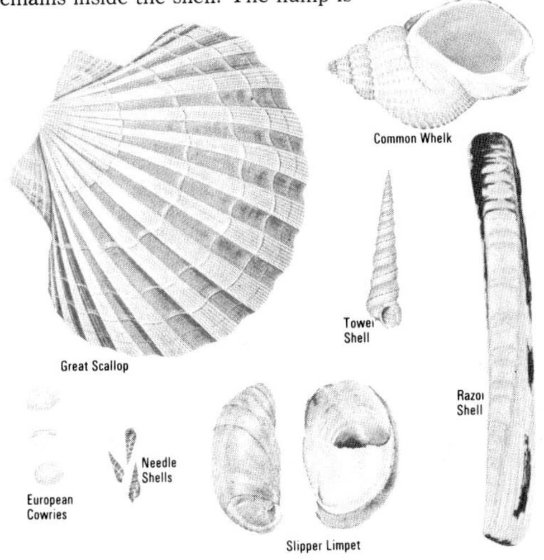

Slugs

Slugs are closely related to snails. They are, in fact, snails which have almost or quite lost their shells. Some are carnivorous and eat earthworms. Many slugs do harm in gardens because they eat plants.

Bivalves

Bivalves are molluscs with two parts to their shells. The two valves are hinged at one edge by little teeth and by an elastic ligament. Muscles hold the shell shut. All bivalves live in water, and most are marine. They include cockles, mussels, scallops, oysters, and clams.

Bivalves have no head. The body consists mainly of the mantle and large gills. All bivalves are filter feeders, sieving tiny particles from the water.

Squids and Octopuses

Squids and octopuses, together with the cuttlefishes and the pearly nautilus, are known as cephalopods. All live in the sea and they make up one of the three main groups of molluscs. Cephalopods are more active than other molluscs. They have a more obvious head than the other molluscs and they certainly have a more efficient brain. They are among the most intelligent of the invertebrate animals. The most obvious features of the cephalopods, however, are the arms or tentacles that surround the head. Octopuses have eight arms, while most other cephalopods have ten. They are equipped with suckers and they are used mainly for catching food.

The cephalopods, like the other molluscs, are soft-bodied creatures and they are surrounded by a thick sheet of skin called the mantle. This normally secretes some sort of shell or skeleton as it does in the snail. The pearly nautilus is the only living cephalopod with an external shell. The red and cream spiral shell is up to 30 centimetres across and it is composed of many chambers. The animal lives in the outer chamber, with its many tentacles exposed.

The shells of other cephalopods are inside the mantle and they are much reduced in connection with the more active life of the animals. The shell of the octopus is almost non-existent, consisting only of a very thin plate or just a few chalky granules. The squid has a thin, horny plate known as a pen, while the cuttlefish has an oval chalky plate known as a cuttlebone. It is often washed up on the beach and it is frequently given to cage birds to provide them with the calcium necessary for the formation of egg shells.

Above: The cuttlefish uses its long arms to catch a shrimp.

Below: Close-up view of mouth of cuttlefish.

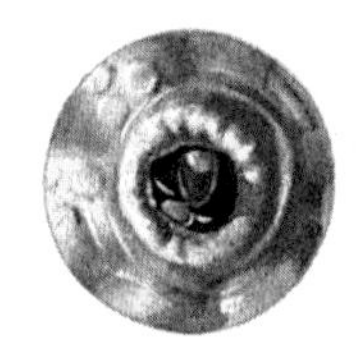

Jet Propulsion

The mantle of the octopus is very muscular and it covers all of the body except the head. It joins the body at the 'neck', where there are a few small openings. When the mantle expands, water is drawn in through these openings and into the mantle cavity. This is a large space around the body in which there hangs a pair of gills. Oxygen is removed from the water and the water is then pumped out again, not through the inlet holes but through

The common octopus is not the dangerous animal that many people imagine. The eyes of these creatures are remarkably similar to our own eyes.

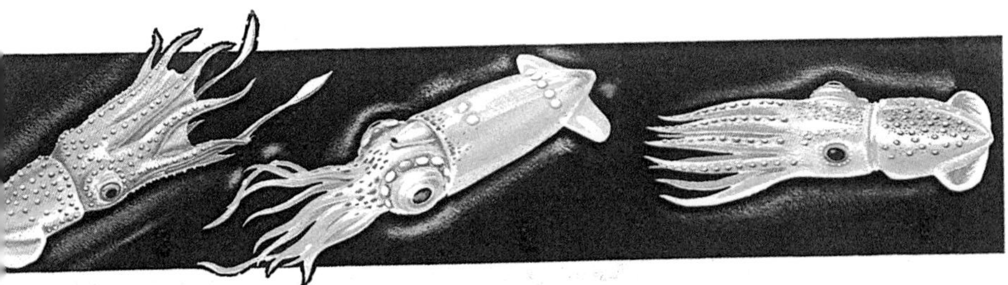

a short funnel called the siphon. The same mechanism is also used for swimming: rapid contraction of the mantle forces a jet of water out through the siphon and the animal shoots backwards.

The squids and cuttlefishes breathe in the same way and they can also jet propel themselves about, controlling their direction by altering the position of the siphon. More usually, however, they swim by gently flapping the fin-like edges of the mantle.

Undeserved Reputation

The octopus is the best known of the cephalopods. It is often portrayed as a fearsome monster, but it does not deserve its reputation. There are about 150 different species and most of them are quite small creatures. An octopus with tentacles spanning 3 metres is a very large one, but even then the body is still quite small. Some octopuses swim freely in the open ocean, but most of them have taken to living on the sea-bed and most of them live around the coasts.

The main food of octopuses consists of crabs and bivalve molluscs. These are caught by the arms and then carried to the mouth. The mouth is in the centre of the eight arms and it has a parrot-like beak which is able to crack the hard shells of the prey. The beak is also provided with poison glands.

The Largest Invertebrates

Squids are much more active than the octopuses. Most of them live out in the open sea and range from the surface layers to the deepest regions. Whereas the octopus is a rather shapeless bag, the squid is a slender creature with a rocket-like shape. There are some 400 species, ranging from a few centimetres to 20 metres in length.

Unlike the octopuses, the squids and cuttlefishes have ten arms. Two of them are much longer and more slender than the others, but they widen out near the tip to form sucker-covered 'hands'. They can be shot out to catch prey. All the cephalopods share the horny beak of the octopus.

Apart from the nautilus, the cephalopods all possess an ink sac. This contains a thick inky fluid which can be shot out into the water when the animal is disturbed. It may form a 'smoke screen' behind which the animal can disappear, or it may simply confuse an attacker for a while.

Above: These three squids live in deep water and they have light-producing organs (see page 91).

Below: The suckers of octopus (bottom), squid (middle) and cuttlefish (top).

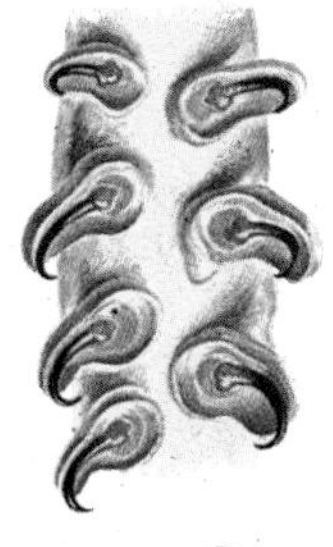

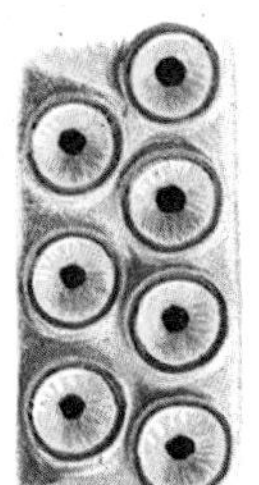

Colour Change Artists

All cephalopods can change their colours to some extent, but the best performers are the cuttlefishes. These animals are rather broader and flatter than the squids and their tentacles are relatively shorter. They live around the coasts and hunt for shrimps and other crustaceans on the sea-bed.

Thunderbolts

In many parts of the world, especially where the rocks consist of Jurassic clays (about 150 million years old), one can pick up smooth stones that look like rifle bullets. Known as thunderbolts, they are fossilized squids called belemnites.

Spiny-Skinned Animals

Together with the brittle stars, sea urchins, and a few others, the starfishes make up the phylum (large group) called the *Echinodermata*. This name means 'spiny-skinned' and refers to the rough skins of the animals. The roughness is caused by numerous chalky plates embedded in the skin. The echinoderms are marine animals of moderate size.

Like the corals and jellyfishes, the echinoderms are radially symmetrical. This means that they are more or less circular or cylindrical creatures. There is no brain nor is there even any structure which we can call a head. Sense organs are poorly developed and special breathing organs are generally absent. Oxygen is absorbed from the water through the skin.

The echinoderms possess a unique feature called the *water vascular system*. This is a system of water-filled canals running to all parts of the body. It is in contact with the outside through a group of little pores on the upper surface of the animal. Tiny branches of the water vascular system grow out through the general body surface of most echinoderms and form *tube feet*. These are little muscular projections with disc-shaped ends. They help to absorb oxygen from the surrounding water and they also enable some of the animals to move about. Water is pumped into the tube feet to force their ends out to make contact with stones or other objects. The muscles of the feet then contract and turn them into little suction cups. These grip very tightly and the animal can drag itself along.

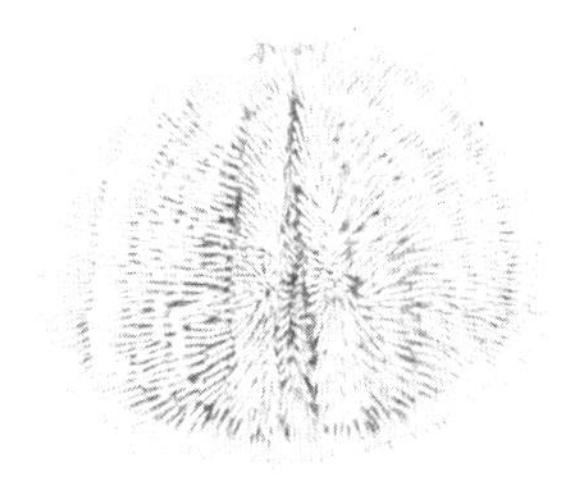

Right: An edible sea urchin.

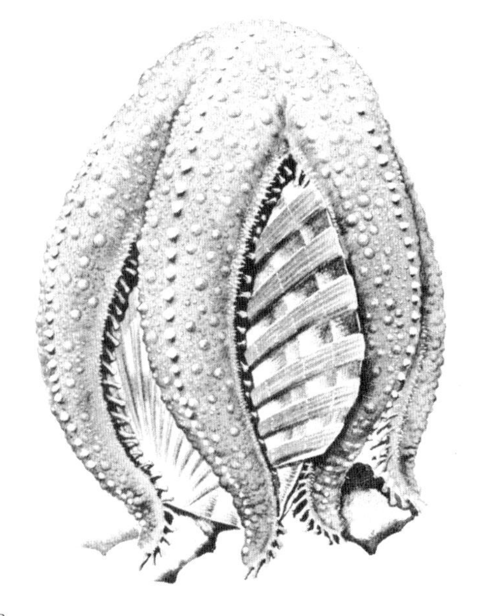

Below: The starfish feeds largely on shellfish. Its tube feet work in unison to exert a strong pull and open the bivalve shell.

Most of the echinoderms have separate male and female individuals, although some are hermaphrodite. Eggs and sperms are usually both released into the water, where fertilization occurs. Young echinoderms are very different from the adults and float freely in the plankton. This helps them spread to new areas.

The Starfishes

Living echinoderms are split into five groups or classes, of which the most familiar is the starfish group. Starfishes are flattened animals and normally have five arms radiating from the central disc. The mouth is on the underside of the body, and so are the tube feet. The latter occur along the

Above: Four kinds of echinoderms. From left to right they are starfish, sea cucumber, brittle star, and feather star.

arms. The skeleton is flexible and is made up of numerous separate chalky plates embedded in the skin. Projections from these plates make the surface rough. Some projections develop into spines; others form tiny pincers. The latter are on movable stalks and they pick pieces of sand and dirt from the skin, thus keeping the animal clean.

Starfishes feed on various molluscs and other invertebrates. They are often serious pests in oyster beds. Some starfishes swallow the molluscs whole and eject the shells later, but others open the shells to get at the soft parts. The starfish's stomach is then pushed out and digestive juices are poured over the mollusc.

Brittle Stars

These animals are often confused with the starfishes, but their arms are much more slender and they are distinctly separated from the central disc. Brittle stars move by coiling their spiny arms around stones and heaving themselves along. The animals often live at great depths and they feed on small creatures. Some brittle stars catch their prey with their arms, others shovel mud into the mouth with their tube feet and digest any creatures it contains.

Sea Urchins

Sea urchins are globular animals whose skeletal plates form a rigid shell called a test. The plates bear many spines and pincers. Tube feet grow out through the test in five distinct rays. The mouth of the sea urchin is on the underside and armed with five strong teeth. The urchin uses these teeth to browse on seaweeds.

Sea Lilies

The feathery arms of sea lilies are attached to the sea-bed by a stalk. They bear tiny cilia which create water currents to carry food particles to the mouth. Feather stars and basket stars are sea lilies which eventually break away from the stalk when they are fully grown.

Above: A sea lily.

Sea Cucumbers

These are sausage-shaped animals in which the skeleton is reduced to scattered little plates. There are five rows of tube feet, and those around the mouth are modified to form tentacles. The tentacles occasionally catch small creatures, but the sea cucumber obtains most of its food by scooping up mud.

Brittle stars differ from starfishes in that the arms are clearly separated from the central disc. Brittle stars heave themselves along by their arms.

Fishes

Fishes are backboned animals (vertebrates) which, with a few exceptions, live entirely in the water. There are more than 20,000 different kinds in the world today, some living in the sea and some living in fresh water. Most of them are beautifully streamlined, and they swim easily through the water by means of their fins and tail. They get their oxygen direct from the water through their gills (see page 98). The gills lie in chambers connected to the back of the throat, and the fishes breathe by taking water in through the mouth.

Backbone
Air Bladder
Kidney
Gills
Heart
Flat Scales
Stomach
Intestine
Ovary
Trunk Muscles

A typical bony fish, showing the internal organs. The swim bladder is shown in blue.

Below: Many tropical fishes are commonly kept in aquaria.

Neon Tetra
Yellow Dwarf Cichlid
Black Widow
Tiger Barb

The fishes were the earliest of the backboned animals to appear on the earth, but the first fishes were very different from those of today. They were heavily armoured and very clumsy. They had no jaws and they lived mainly on the bottom, sucking up mud and filtering out small animals and decaying material. Fishes with jaws appeared later and few of the jawless fishes are still alive today. They are known as lampreys and hag-fishes.

Below: Not many fishes look after their eggs or young, but when danger threatens Tilapia gathers its young into its mouth.

Cartilage Skeletons

Apart from the jawless lampreys and hag-fishes, today's fishes fall into two main groups: the cartilaginous fishes, whose skeletons are made of gristly cartilage, and the bony fishes. Both groups possess two pairs of fins on the sides of the body. The cartilaginous fishes live in the sea and they include sharks, rays, skates, and dogfishes. Most of the cartilaginous fishes can be distinguished from bony fishes by looking for the gill openings. The sharks and their relatives have five gill slits on each side, but the gill slits of the

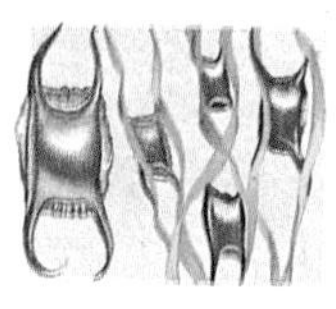

Above: Many shark-like fishes, including the skate and the dogfish, lay their eggs in horny cases known as mermaids' purses.

Thick-lipped Mullet
Cod
Rudd
Wrasse
Dace
Mackerel
Salmon
Norway Haddock
Perch
Bass
Barbel
Sailfish
Blue Shark
Plaice
Male Enlarged
Sting Ray
Deep Sea Angler Fish
with Parasitic Male

bony fishes are hidden under a flap called the operculum.

Some of the sharks grow to a great size. The whale shark is the largest of all living fishes. It reaches lengths of more than 7 metres. Cartilaginous fishes have very rough skins, which can be used as sand-paper. The roughness is due to the tooth-like scales that cover the skin. A shark's teeth, carried on the skin of the lips, are simply enlarged and razor-sharp versions of the normal body scales. Most sharks feed on other fishes, but the skates and rays are bottom feeders and prefer shell-fish. Instead of sharp cutting teeth, skates and rays have blunt teeth for cracking and crushing mollusc shells.

Bony Skeletons

Nearly all of today's fishes are bony fishes, with skeletons of real bone. Examples include herring, cod, perch, stickleback, salmon, and plaice. Some have no scales. The body is generally flattened from side to side and much more slender than that of the sharks.

Fishes with Lungs

There are two main groups of bony fishes. One group possesses fins of fine bones or rays. The other group has muscular fins, with sturdy bones in the centre. The ray-finned fishes breathe with their gills and their air bladders have developed into buoyancy tanks. The fishes with muscular fins can also crawl out of the water and breathe by gulping air into their air bladders. The ancestors of these fishes gradually evolved into land animals but a few have remained fish-like. They are known as lung fishes.

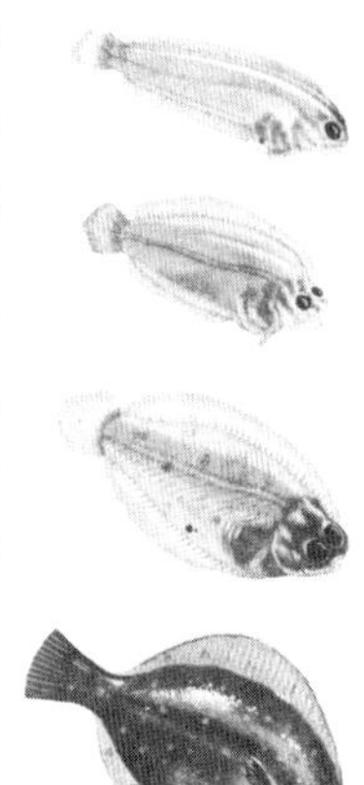

Above: The true flat-fishes, such as the plaice, lie on one side. During the fish's growth one eye moves round so that both are on the top side. The head becomes distorted.

The seahorse, a very unusual fish.

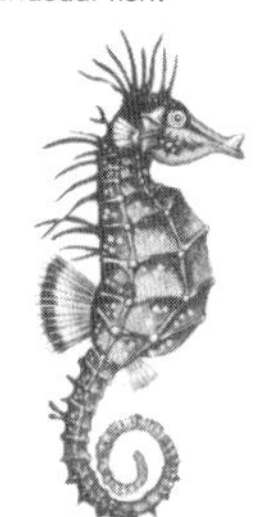

Below: The bichir is a primitive bony fish with lungs, although not related to the true lung fishes.

Apart from the various bottom-living species, all bony fishes have a swim bladder, or air bladder. This is a white or silvery pouch lying just above the food canal. In some of the more primitive fishes, the swim bladder has a duct connecting it with the throat, but in most bony fishes, however, it is completely closed. It contains oxygen and it acts as a sort of buoyancy tank, enabling the fish to float effortlessly at any level in the water.

Those fishes whose swim bladders are connected to their throats can gulp air into them. Some can live for a while out of water because they can breathe in this way.

Reproduction in Fishes

The majority of fishes lay eggs, and some of them lay enormous numbers. A female cod may contain up to nine million eggs. Not all fishes lay vast numbers of eggs. In general, the number of eggs depends upon the degree of care that the eggs and young fishes receive from their parents. The cod simply scatters her eggs in the water and leaves them. She has to produce large numbers to ensure that at least some escape being eaten. At the other end of the scale, the little freshwater bullhead lays only a few dozen eggs. These are deposited in a hollow scooped out under a stone on the stream bed, and the male keeps a watch over them until the young fishes are several days old.

The stickleback is another fish that cares for its eggs and young. The male makes a nest with water plants and twigs and he encourages one or more females to lay their eggs in it. After the eggs are laid the male stands guard and ensures that they get plenty of oxygen by fanning a current of water over them with his tail. He does the same for the young fishes.

The seahorses have an even more unusual breeding behaviour. The male has a pouch on his belly and the female lays her eggs in it.

Amphibians

The amphibians were the first back-boned animals to leave the water and live on land. That happened some 400 million years ago, but even today the amphibians have not completely broken away from the water. They have moist skin and they thrive only in damp places. Nearly all have to go back to the water to breed. The name amphibian means 'double life' and indicates that the animals live partly on the land and partly in the water. The life history of the common frog is typical. The young stages, known as tadpoles, breathe with gills. They gradually change into the air-breathing adults.

There are two main groups of amphibians living today. These are the frogs and toads on the one hand and the tailed amphibians (newts and salamanders) on the other. There is a third group, found in the warmer parts of the world. They look like large earthworms, and they spend their lives burrowing in the soil.

Frogs and Toads

These are jumping amphibians. They use their long, webbed hind legs for both jumping and swimming. We tend to call the smooth-skinned members of the group frogs and the rougher skinned ones toads, but this is not a very scientific division.

Some frogs and toads live permanently in the water, especially in swiftly flowing streams. They breathe entirely through their skin. Some other frogs and toads live entirely on the land. They lay eggs in damp soil and the tadpole stage is passed inside the egg. When the animals finally come out of their eggs they are already small frogs. The majority of frogs and toads, however, lead lives like that of the common frog. The tadpoles live in water, but the adults spend most of their time on the land and return to the water only for a short time during the breeding season. Some tree frogs make use of the small 'ponds' formed in the forks of trees or between the leaves of other plants.

Most frogs and toads leave their eggs to look after themselves, but a few species show some concern for the welfare of their young. The European midwife toad mates and lays its eggs on land. The male carries the spawn about wrapped around his back legs. When they are ready to hatch, the male enters the water and releases the eggs. A few frogs and toads build nests for their eggs. The male Surinam toad lays its eggs onto the female's back and they sink into small hollows in her skin. The skin grows over to form a lid. The tadpoles complete their development inside these pouches.

Toads usually have a thicker, rougher skin than frogs and their back legs are shorter. Toads cannot jump as well as frogs and often crawl about.

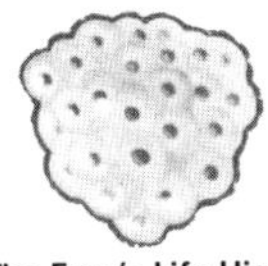

The Frog's Life History

Adult frogs hibernate during the winter, hidden under logs or stones or buried in mud. In spring the males return to the water. Their loud croaking attracts the females, and the animals pair up in the water. The eggs are covered with jelly which swells up as soon as they get into the water and protects them.

After a few days, the black egg develops a distinct head and tail. The tadpole emerges from the jelly after 10 days. It is three days before the tadpole can have its first meal of plant scrapings. It has now three pairs of feathery external gills. During the next few weeks the external gills gradually disappear and they are replaced by internal gills. The hind legs start to form when the tadpole is about six weeks old, and the front legs appear a few weeks later. The diet has now changed from plant material to small water creatures. Lungs have formed inside the body by this time. The legs continue to grow and the tail begins to shrink. The animal takes on the typical frog shape.

The frogs and toads that live permanently in the water have lost their tongues and they grab their food with their jaws. The land-living species normally have long tongues which are sticky at the tip. They are flicked out to pick up a fly or some other small creature.

Newts and Salamanders

Most of the tailed amphibians are called salamanders. They vary from a few centimetres to about 1.8 metres in length. Newts are only a particular group of salamanders. The eggs are fertilized inside the female's body and she usually lays them separately on water weeds. Otherwise the life history is very similar to that of the frog, although the newt tadpole keeps its external gills until it becomes an adult.

A frog showing the moist glistening skin which is characteristic of these animals.

Below: The midwife toad, a European species in which the male looks after the eggs.

The European spotted salamander is one of many viviparous species. This means that it brings forth active young. The eggs hatch while still in the mother's body. Some other salamanders never go into the water. Some of them lay eggs in damp soil and miniature adults emerge from the eggs. The black salamander of the Alps takes the process even further. The eggs hatch while still in the mother's body, but the young are not born until they have passed through the tadpole stage and turned into tiny adults.

Salamanders that do not grow up

Many salamanders spend all their lives in the water. Quite a number of them have no lungs and some retain their feathery external gills all their lives. They look like overgrown tadpoles, although they grow up in other ways and can reproduce themselves. One of the best known is the axolotl. It is the tadpole of a Mexican salamander.

Above: Male newts have bright colours and large crests in the breeding season. Below: The fire salamander and a caecilian.

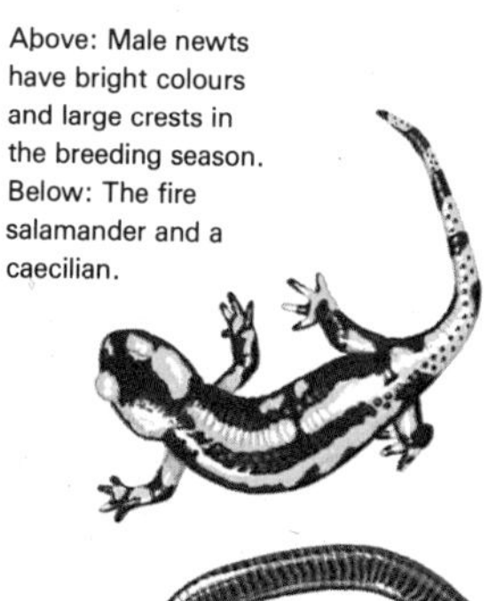

Reptiles

Reptiles are cold-blooded vertebrates with dry, scaly skin. Most reptiles live in the warmer parts of the world where their cold-bloodedness is not much of a handicap. There are some in the temperate regions, and the adder reaches the Arctic Circle.

The reptiles evolved from the amphibians. The two most important features which separate them are the waterproof skins and the tough shells of the eggs. These mean that reptiles are independent of the water.

Living reptiles fall into three main groups: the tortoises and turtles, the crocodiles and alligators, and the lizards and snakes. A fourth group is represented only by the New Zealand tuatara (see page124).

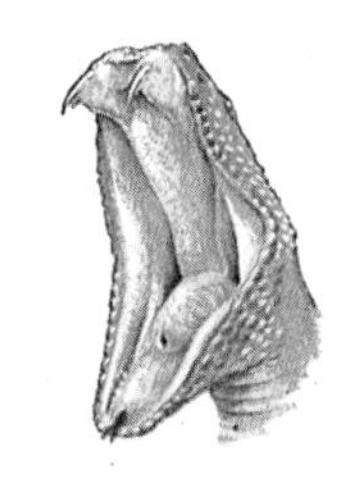

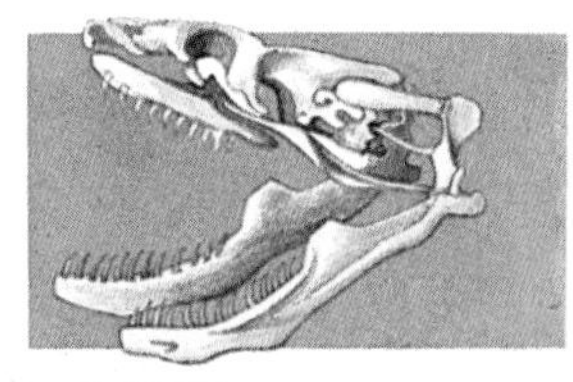

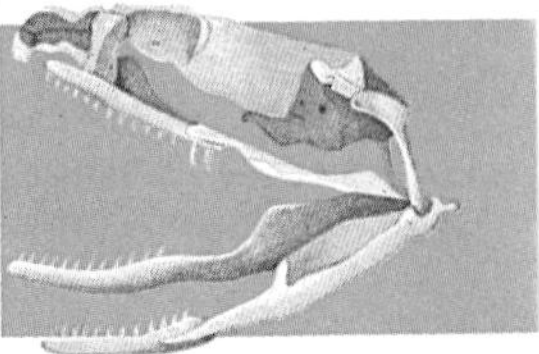

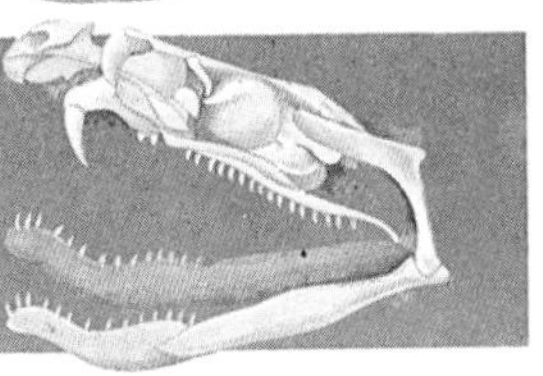

Non-poisonous snakes, such as grass snakes and boas, have no fangs (top right). Some snakes have fangs at the back of the mouth (centre) but these are not usually very dangerous. The most dangerous snakes are the front fanged species, such as the cobra (right). Viper fangs (above) fold away when the snake shuts its mouth.

Tortoises and Turtles

There is rather a muddle over the names of some of these animals. An individual may be called a tortoise in one country, a terrapin in another, and a turtle in a third. But there is a tendency to use the name tortoise for those living on land, terrapin for fresh-water species, and turtle for marine species.

Left: The thorny devil, an Australian lizard.

The animals are easily recognized by their box-like shells, made of bone and normally covered by horny plates. There are gaps for the limbs and the neck, and the head can usually be withdrawn into the shell by bending the neck.

Land-dwelling tortoises feed mainly on plants, but water-dwelling species are mainly carnivorous. None of them has any teeth, but the jaws have a sharp, horny beak.

All tortoises and turtles lay hard-shelled eggs. The marine turtles come ashore to lay their eggs on a sandy beach.

Crocodiles and Alligators

These are the closest living relatives of the great dinosaurs. They live mainly in tropical rivers, although some species venture into the sea. They spend most of their time basking on the river bank or floating just under the surface of the water. The eyes and

Many animals trick their enemies by pretending to be something else (see page 88). This frilled lizard raises a collar of skin around its neck, making itself look larger and fiercer than it really is.

Left: The egg-eating snake feeds on birds' eggs which it breaks with special 'teeth' in its throat.
Right: The cobra raises a hood of skin when alarmed.

nostrils are on the top of the snout and they just break the surface when the animal is floating. The animals swim by moving their powerful tails from side to side.

Crocodiles and alligators feed on a variety of other animals. Fishes are their principal food, especially when they are young, but they also catch birds and mammals which come down to the water to drink.

Crocodiles and alligators look very much alike, but they can be distinguished by their teeth. There are only two kinds of alligator, one in the southern part of the United States and one in China. There are many different kinds of crocodiles. Other members of the group include the caimans, closely related to the alligators, and the gavial. The latter animal has a very long, narrow snout and it feeds on fishes.

Some crocodiles lay their eggs in pits, but others, together with the alligators, lay them in heaps of rotting vegetation. The heat of fermentation helps to incubate the eggs. The mother stays near the nest and, when she hears the young animals breaking out of their eggs, she removes the covering material and helps them out.

Lizards

The lizards are a very varied group of reptiles. Most are only a few inches long. The Komodo dragon, however, reaches a length of 4 metres. Most lizards have four limbs, but some species, the slow worm for instance, are legless and look very much like snakes.

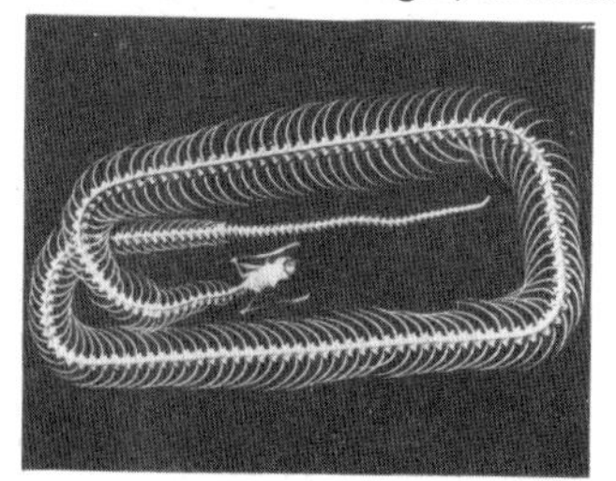

Nearly all lizards feed on animal material, such as insects. The chameleons are famous for catching flies by shooting out their sticky tongues with remarkable speed and accuracy.

Most lizards lay eggs, but a few species bring forth active young. The common lizard of Europe is one.

Snakes

Snakes are legless reptiles. Not all are harmful. In fact, only about 150 out of over 2,000 species are at all dangerous to man.

Snakes have two main ways of moving. Many have broad scales on the belly which can be lifted and dug into the ground so that the snake can pull itself forward. Other snakes throw their bodies into loops and they glide forward by pushing against the ground.

Constricting snakes wrap their bodies around their prey to suffocate it. Other snakes catch their prey by biting. The poisonous snakes have hollow or grooved teeth called fangs, which carry poison into the wound. The prey is swallowed whole.

Most snakes lay leathery eggs, but some species bring forth active young.

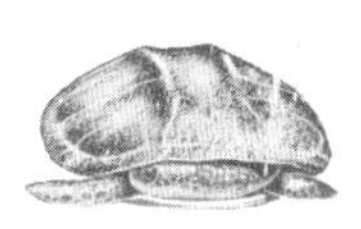

Above: Turtles can withdraw their heads into their shells by bending the neck sideways (top) or vertically. The shell has played a large part in their survival.

Below: Snakes have very many vertebrae and joints in their backbones.

Birds

Birds are the only animals with feathers. They are warm-blooded and the feathers help to keep them warm. The feathers also enable birds to fly. Birds probably descended from some kind of tree-living reptile which used to glide from branch to branch.

There are nearly 9,000 species of birds living today, with a very wide range of sizes, shapes, and habits. Many birds have now lost the ability to fly. They include ostriches, emus, kiwis, and penguins.

Ostriches and emus are large running birds. The ostrich is the largest living bird. A male may reach a height of about 3 metres and a weight of about 170 kilograms. The elephant bird, which lived in Madagascar, weighed nearly half a ton. The tallest birds were the moas of New Zealand which reached 4.7 metres in height. The heaviest flying bird alive today is the mute swan. It reaches 28 kilograms in weight. The smallest bird is the bee hummingbird of Cuba. It has a total length of 5 centimetres and weighs 30 grams.

Above: The various stages in wing movements of a pigeon.

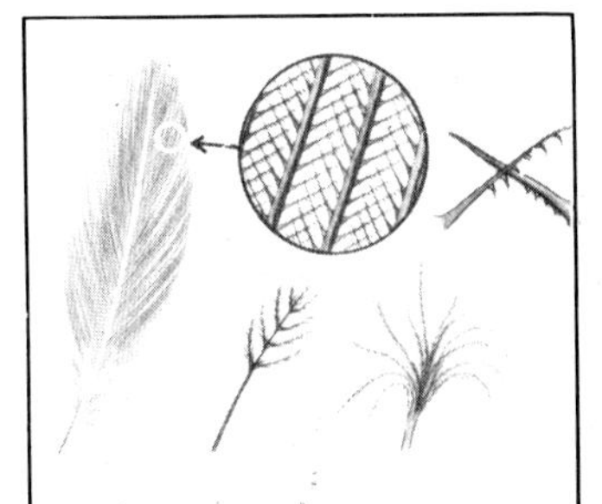

Feathers

There are several different kinds of feathers. The *contour* feathers cover the bird. They have a stiff, horny shaft and a flat part called the *vane*, consisting of hundreds of tiny branches called *barbs*.

How Birds Fly

Any flying object, whether a bird or an aeroplane, must have *thrust* to push it along and *lift* to keep it in the air. Some birds glide and soar almost effortlessly on air currents, but most birds fly by flapping their wings. In flapping flight the wings provide both thrust and lift. Thrust is provided by the twisting movement of the feathers at the wing tips. This movement forces the air backwards and therefore drives the bird forwards. The feathers also form the curved, or cambered, upper surface which is responsible for the lift.

The greater the area of a wing, the more lift it can provide. Large birds, such as swans, therefore have large wings to lift them. Narrow wings are best for fast flight and for birds such as albatrosses that glide through fast-moving air. Small birds, with little weight, have short wings. Vultures, buzzards, and other birds that fly slowly or soar in circles have very broad wings. These provide plenty of lift at slow speeds.

Feeding

Birds use up a great deal of energy when flying. They also need a lot of energy to keep their bodies at a temperature of about 40°C. This means that birds need a lot of food. The horny beak is used for catching and cutting up food. Birds have a wide variety of beaks, each suited to the diet. Examples are hooked beaks of hunting birds, stout beaks for cracking seeds, and slender sharp beaks for catching insects.

Far right: Nests of weaver finches—masses of interwoven grasses providing safe shelter.

Sight and Hearing

The sense of smell is poorly developed and birds rely on the senses of sight and hearing. Sight is extremely good in birds. Hawks hovering high in the sky can spot a tiny mouse in the grass below and swoop down to catch it. Kingfishers on their perches can see the outlines of fishes in fast moving water and dive down to get them. Barn owls pounce on mice in pitch darkness by homing in on their rustling noises.

Nest Building

At the start of the breeding season the males of many species adopt territories. They stake their claims by singing and displaying, defending the territory against other males, and attracting females. Nests are used only during the breeding season, although some birds return year after year to the same nest. Some species make do with a rocky ledge or with a hollow in the ground, but others make very elaborate nests. The birds collect twigs, grass, moss, hair, mud, and other materials and work it together into a cup-shaped nest.

When the nest is complete the eggs are laid. Some species lay only one egg but others lay as many as twenty eggs. The eggs have to be incubated—that is, they have to be kept warm. The parent birds keep them warm by sitting on them. Sometimes both parents take it in turns to sit on the eggs, sometimes only one parent sits. When the eggs hatch the chicks need plenty of food. The parents are very busy at this time, often bringing food to the nest four times every minute.

Some young birds, especially those nesting on the ground, are fully feathered when they hatch and they are able to run about right away. Most tree-nesting birds, however, are blind and naked when they hatch. The parents continue to keep them warm. Small birds probably do not live more than four or five years in the wild. Larger birds live longer.

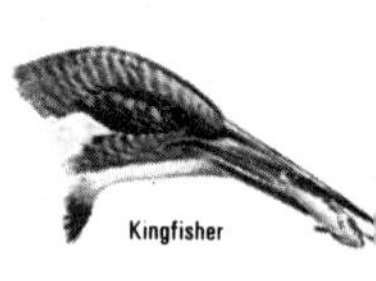

There is a great variety of bird beaks. They include the crushing beak of the macaw, the sifting beak of the shoveller, the seed-cracking beak of the bullfinch, the dagger beak of the kingfisher, and the sharp, hooked beak of the falcon.

Mammals

The mammals are the dominant animals in the world today. They have a marked effect on other animals and on the land in general. One of the main factors which have helped the mammals to dominate the earth has been their intelligence. Mammals have large brains and they are able to learn quickly.

Mammals are easy to recognize by their fur coats. All mammals have some hair, although there is not much of it on an elephant and a whale only has a few bristles. The feeding of the young on milk from the mother's body is unique. There are also various internal features which separate the mammals from the other vertebrates.

Left: The echidna or spiny ant-eater, one of the monotremes or egg-laying mammals.

For example, the lower jaw of a mammal consists of only a single bone on each side. Experts use this fact to decide whether fossil bones come from mammals.

The mammals came to prominence quite recently in the earth's history, about 70 million years ago, when the dinosaurs became extinct. There are less than 5,000 species alive today and several of the orders of mammals are on the decline.

There are three major groups of living mammals: the *monotremes*, the *marsupials*, and the *placental mammals*. The first two groups are almost confined to the Australian region, and the world is ruled by the placental mammals, to which we belong.

The Monotremes—mammals that lay eggs

The mammals originally descended from egg-laying reptiles. Most of the mammals gave up the egg-laying habit, but one small group broke away from the others very early in their history. They did not give up laying eggs, and they also kept many other reptile features. A few of these primitive mammals are still alive today. They are the monotremes which include the duck-billed platypus and the spiny ant-eaters. The latter are also called echidnas. Although they lay eggs, they still look after their young and feed them on milk, so they are true mammals. They also have hair and warm blood, although they are not so good at keeping their temperature steady as the other mammals. The monotremes are found only in Australia and New Guinea.

Right: Prairie dogs are American members of the squirrel family, part of the largest of all mammal orders—the rodents.

Right: The American fieldmouse is one of the smaller members of the rodent order.

The Marsupials—mammals with pouches

The marsupials include such well known animals as the kangaroos and the koala bear. The young are born at a very early stage of development. The red kangaroo may be over two metres tall, but it gives birth to a tiny baby only two centimetres long. This tiny baby is quite unable to look after itself, but it manages to crawl into its mother's pouch. This is a large fold of skin on the mother's belly and her milk glands open into it. The baby stays in the pouch for some months and feeds on the milk. Eventually it is sufficiently well developed to leave the pouch and hop about, but it still returns to its mother's pouch from time to time until it is quite able to look after itself. Other marsupials start life in the same way, although some of the opossums have no real pouch.

Most marsupials live in the Australian region, although some live in America. At one time they were found all over the world, but the placental mammals gradually replaced them in most places. Australia and South America had become separated from the rest of the world before many of the placentals reached them. The marsupials thrived there for a time. South America was later connected to North America and placentals from the north wiped out most of the marsupials. A few opossums survived, however, and the Virginia opossum even managed to invade North America.

Marsupials are still quite numerous in Australia, but their numbers have dropped a great deal since Europeans introduced cats, dogs, foxes, and other placental mammals. Several species, especially the smaller ones, have become extinct.

The Placental Mammals

This is the most advanced and most intelligent group of mammals. The female keeps her babies inside her body for a relatively long period and she feeds them through a special organ called the *placenta.* This is attached to the mother's womb and the baby is connected to it by the *umbilical cord,* which carries blood vessels from the baby to the placenta. In the placenta, the blood vessels from the baby lie very close to those of the mother. Food and oxygen can then pass from the

Below: Representatives of each of the orders of mammals. (Not drawn to scale.)

Bush-babies (above) and vervet monkeys (above right) are two representatives of the order primates. Man is classified in this order.

mother's blood into that of her baby.

Placental babies are much more highly developed than the babies of marsupials, but many of them are still quite helpless. The young of dogs, cats, rabbits, and many others are blind and naked when born. Their mothers have to do everything for them. The young of horses, antelopes, and other animals are able to get up and run about as soon as they are born. During the time they are with their mother they learn a great deal about hunting or finding food. This period of learning while in the care of the mother (and father as well with some animals) has surely contributed to the success of the placental mammals.

Rhinoceroses are hoofed mammals belonging to the same group as the horse and tapir. There are five species, but only the two African species—the black rhino and the white rhino—are at all numerous today. Even these number only a few thousand.

Instinct and Learning

The thrush does not have to be taught how to build its nest, and nor does the spider have to be shown how to make its web. The animals know automatically how to perform these tasks. Behaviour patterns of this kind are called *instincts*.

Instincts account for a great deal of animal behaviour, including reaction to danger, courtship displays, rearing and feeding of the young, and migration. The behaviour pattern needs something to trigger it off. The trigger is called a *releaser*.

Each kind of animal has its own types of instinctive behaviour, and all members of a species will normally behave in the same way. The instincts are inherited and behaviour patterns have evolved in just the same way as colours and shapes.

Some instincts are very simple actions. Woodlice, for example, scuttle about in an agitated fashion if exposed to bright light or dry air. They keep this up until they reach a dark or damp place and then they settle down. This is a simple action, but it ensures that the animals keep themselves in the right conditions. Other instincts are much more complicated. Nest-building, for example, involves collecting various kinds of materials and joining them together to make the layers of the nest. Even more complex are the engineering feats of the beaver. This animal knows instinctively how to fell trees and cut them into logs. He also knows how to dig canals and float his logs along them. He then uses the logs to build dams which raise the water level and allow him to build a safe home in the middle of the pond. Yet all this is instinctive behaviour and the beaver is not clever.

Courtship behaviour is a very complicated business for many animals. Pheasants and peacocks display their fine feathers to the females. Male redstarts hold singing contests to impress

Animals spend more time threatening their opponents than actually fighting, like the show of strength by this elephant.

Below left: Blue tits soon learn to peck through milk tops to get at the cream, although this is a new habit for the species.

Below: The courtship of the whooping crane.

the hens. Male fishes often perform elaborate 'dances' to engage the attention of females. All this is instinctive behaviour. There are usually several stages, and each stage normally requires a separate trigger. In courtship behaviour, for example, the female's reaction to one stage will trigger off the next stage in the male's display.

Some instincts are perfect from the start. For example, a silk moth only makes one cocoon in its life, but it is perfectly made. Other instincts have to be perfected by practice. Birds know how to fly by instinct, but they need some practice before they master it.

Because instinctive behaviour is triggered off automatically, by a simple releaser, animals can often be misled. A male robin will attack a bundle of red feathers from another robin, but it will ignore a stuffed robin whose breast feathers have been painted brown. The attacking behaviour is triggered off simply by sight of the red breast feathers.

Above: The male lyre bird displays his fine plumage to the female when courting. This display triggers off the female's mating behaviour.

Left: These baby shrikes have just left the nest. They know how to fly, but they will have to have several practice flights.

Right: The woodpecker finch of the Galapagos Islands has developed the habit of using a thorn to dig insects out of crevices.

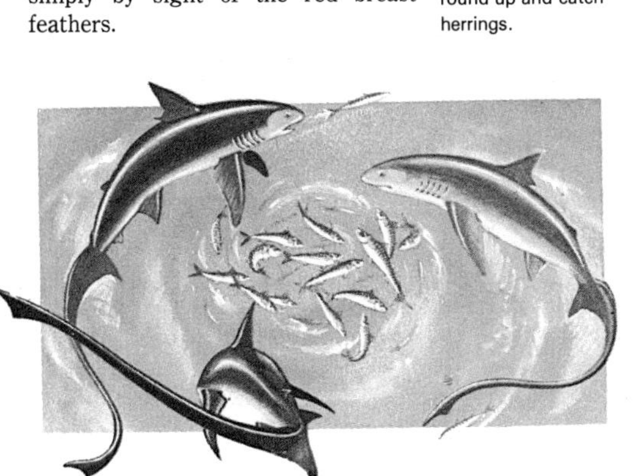

Below: Thresher sharks co-operate to round up and catch herrings.

Learning

The lives of the lower animals are ruled almost entirely by instinct. Mammals, birds, and some other animals are able to learn to some extent. They remember things that have happened before and modify their instinctive reactions so that they can fit their behaviour to a new set of circumstances. Much of the behaviour of birds and mammals is probably a combination of instinct and learning, with learning more important in mammals.

Instincts and Hormones

Some instincts appear only at certain times. They are often due to the action of hormones. These are chemicals released into the blood stream. They have marked effects on behaviour but they do not actually cause behaviour patterns to begin. Hormone activity is high during the breeding season and a cock bird will display vigorously when he sees a female. Out of the breeding season, when the hormones are not active, courtship behaviour ceases.

Animal Parasites

Parasites live in close association with other animals, called the hosts, and they take food from them. They do not usually kill the hosts, but they do weaken them.

Internal parasites live inside the bodies of their hosts. The most important members of this group are tapeworms, roundworms, and flukes. Some live in the blood system of the host animal, but the majority live in the intestine. Some worms merely absorb the host's digested food. Worm-infested animals are therefore always thin and hungry. Other kinds of worm actually eat the host's tissues.

Internal parasites do not move much and they have reduced muscles and sense organs. They concentrate on reproducing and lay huge numbers of eggs, but very few reach the right host and grow into adults.

Right: These strange objects growing on a rose leaf are galls caused by the grubs of a tiny gall-wasp.

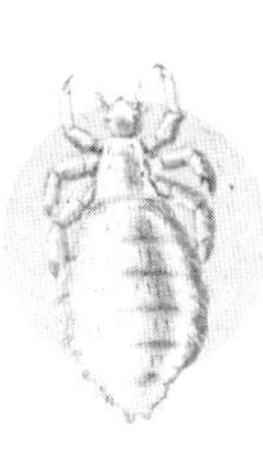

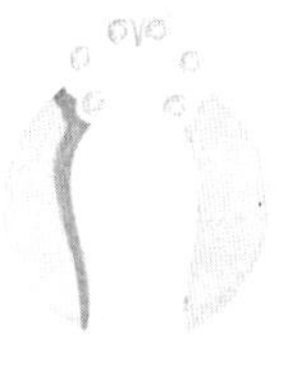

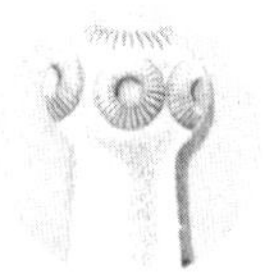

Above: Parasites attach themselves firmly to the host. The louse (top) has claws; the fluke (middle) has suckers; and the tapeworm (bottom) has both hooks and suckers.

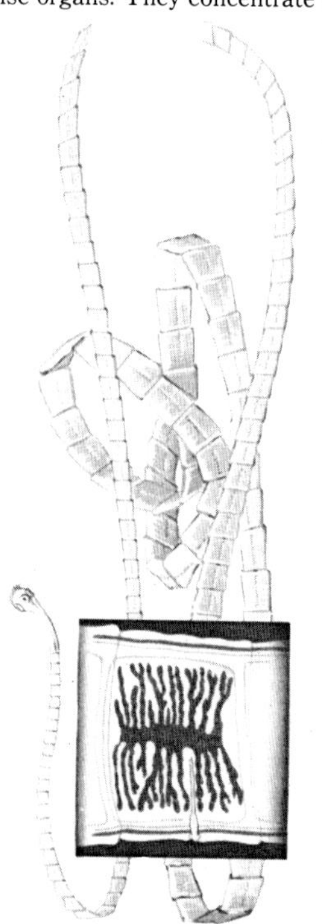

Left: The tapeworm's body consists of a small head and a string of segments containing mainly reproductive organs.

Oak Apples and Other Plant Galls

A soft pinkish outgrowth is a common sight on oak twigs in the summer. This is the oak apple. It is not the fruit of the oak tree, but an abnormal growth caused by an insect. If you cut it open you will see a number of small grubs inside. In the spring a small brown gall-wasp laid its eggs in the oak twig, and when the grubs hatched out they caused the twig to swell. Galls like this are quite common on many kinds of plants. Most are caused by insects, but some are caused by mites.

External parasites live on the outside of their hosts and most of them suck blood. Examples include ticks, lice, and fleas. Most have strong claws or hooks to hang on to the fur or feathers of the host, and many carry disease germs. Fleas can carry plague, lice can carry typhus fever, and ticks can carry many serious diseases of cattle.

Animal Partnerships

As well as the one-sided association of a parasite and its host, there are associations in which both partners derive benefit from living together. This is called *symbiosis*, which means 'living together'.

Animal Partnerships

One of the best known examples of symbiosis involves a hermit crab and a sea anemone. Hermit crabs live in old seashells. They have to change to larger ones to accommodate them as they grow. A sea anemone sometimes grows on the shell and gives the crab protection from its stinging cells, while feeding on scraps from the crab's meals.

Oxpeckers are small birds which run about on the backs of African animals, such as buffaloes, rhinoceroses, and antelopes. Far from annoying the animals, they perform a useful service by eating ticks and other bloodsuckers. The birds warn their hosts of approaching danger.

Some animals cannot live without their symbiotic partners. Very few animals can produce the necessary juices to digest wood. Termites have hordes of protozoans living in their stomachs which break down the wood, in return for shelter and keeping a proportion of the food for themselves.

Plant Partnerships

The partnership between algae and fungi in a lichen is described on page 20.

The roots of clover and other members of the pea family bear swellings called root nodules. These contain bacteria which convert atmospheric nitrogen into nitrates. Nitrates are essential for plant growth and the bacteria get shelter and other food materials from the plants.

Mixed Partnerships

Partnerships between plants and animals are quite common. Corals contain green algae in their bodies. Their main role is probably to use up waste materials produced by the animals.

Most grazing animals keep large populations of bacteria in their digestive systems to break down the cellulose of plants and produce sugars for the animals to absorb.

The cattle egrets on these buffaloes feed on the insects that the animals disturb. In return, they warn the buffaloes of approaching danger.

Survival

For animals which can fall prey to larger animals, survival may depend on having a way of defending themselves.

Above: The colouring of many animals blends well with their natural habitats.

Camouflage

Camouflage helps animals to hide from their enemies but it also helps the hunting animals, such as leopard, to stalk up on their prey. In *counter-shading* the underside of the body is generally somewhat lighter in colour than the upperside, thus tending to cancel out the shadows. This helps the animal to merge with the background. Irregular stripes and patterns are very effective at camouflaging animals because they break up its outline. The tiger and many fishes use this method of camouflage.

Some animals can change colour to match the background. The chameleon is the best known of these creatures, although it is not so good at changing its colour as some animals. The cuttlefish can change colour faster than any other animal. The skin of the cuttlefish (and of its cousins the squids)

The pale wing tips and thorax of the buff-tip moth (below) look just like broken twigs when the insect is at rest (left).

Below: Many young deer are marked with white spots. These help to conceal the animals in the woodlands.

has three layers of pigment cells. One layer is black, one is red, and the outer is yellow. Each pigment cell is controlled by a nerve and it can contract or expand very quickly. The animal can produce a whole range of colours from black, through reds and browns, to yellows according to which cells are expanded. Because each cell has its own nerve the animal can expand the cells in one part of the body and not in another, producing a variety of patterns.

Many fishes, such as the plaice, can change colour to match either sandy or stony backgrounds. In experimental work some of these fishes have even adopted a chequered pattern matching a chess board.

In their behaviour and appearance the hoverfly (right) and the wasp beetle (below right) mimic certain wasps. They are left alone by birds and other predators although they are harmless themselves.

Below: The eyed hawkmoth is quite inconspicuous as it rests on a tree trunk. When disturbed, however, it raises its front wings and displays two large eye spots.

Animal Twigs

A number of animals, especially among the insects, have developed a similarity to objects in their surroundings. The majority resemble leaves and twigs.

Spines and Prickles

Possession of spines is a common form of defençe. Hedgehogs roll up into very prickly balls and they are then safe from most of their enemies. Spiny fishes, such as the porcupine fish

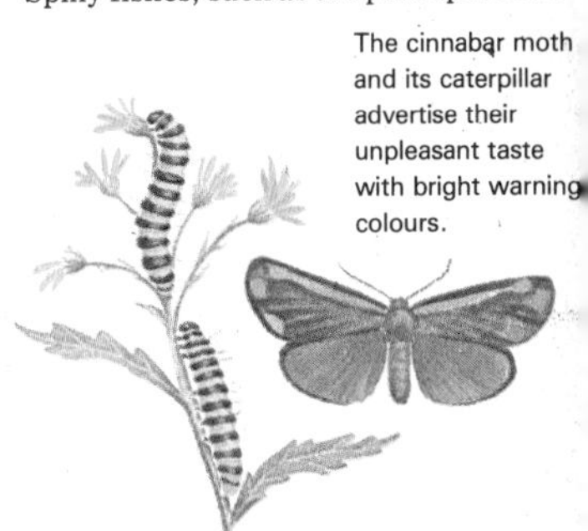

The cinnabar moth and its caterpillar advertise their unpleasant taste with bright warning colours.

and the pufferfish, inflate themselves with water when attacked. Other animals may have a more active form of defence in the possession of stings. Bees, wasps, and scorpions are good examples. A further group of animals, including such diverse creatures as the skunks and ladybirds, protect themselves by a nasty taste or smell.

Above: Prawns become lighter or darker, according to the background, by changing the size of the red pigment cells.

Right: Some animals with stings or poisonous bites are marked with bold *warning colours*.

The red underwing caterpillar is almost invisible as it lies along a twig. Shadows are obliterated by a fringe of hairs along the underside.

Warning Coloration

Animals with nasty tastes or stings are often marked with bold colours and they are very conspicuous. For example, many wasps are clearly marked with black and yellow. These bold colours are *warning colours*, because they warn birds not to touch the insects.

Mimicry

The bold colours of bees and wasps are mimicked by a variety of harmless insects, which fool birds into thinking that they, too, are harmful. These include several different kinds of hoverflies, beetles, and moths. The bees and wasps are the *models*, and the other insects are the *mimics*.

The porcupine is adequately protected by its barbed quills.

Another kind of mimicry involves two or more inedible or otherwise unpleasant species sharing the same warning pattern. An insect-eating bird therefore has only one pattern to learn before it avoids all the insects sharing it.

Eye Spots for Protection

Small eye spots near the tail of a fish or on the wings of a butterfly deceive an enemy and cause it to strike at them. The animal then escapes. Larger eye spots are used to frighten enemies because they look like eyes.

Hibernation

Many animals living in the cold and temperate regions disappear in the autumn because they cannot withstand the cold, or because they are unable to find food during the cold season. Birds usually migrate to warmer lands, but many other animals go into *hibernation*. This is a very deep sleep, during which the body processes slow down almost to a stop. Very little energy is consumed and the body temperature, even in mammals, falls to within a degree or two of that of their immediate surroundings.

Many invertebrates pass the cold season in the egg stage and their eggs are extremely resistant to the cold. Many butterflies and moths pass the winter in the pupa stage or chrysalis. Some species hibernate as caterpillars, and these can sometimes be found among the dead leaves in ditches and hedgerows. A few butterflies and moths hibernate as adults. When the days grow cold they find some dry corner—perhaps in a hollow tree or a garden shed—and go to sleep until the first warm days of spring. Several species of snail hibernate under stones and logs. They seal up their shells with mucus.

Frogs, tortoises, snakes, and lizards bury themselves away from the effect of frost. They often huddle together and this habit undoubtedly helps to keep their temperature a degree or two above that of their surroundings. Fishes do not really hibernate, although some species become lethargic and partly bury themselves in the mud.

Hibernation in Warm-blooded Animals

Some birds become drowsy in cold weather, but true hibernation is not known among birds. The nearest approach to hibernation is shown by the whip-poor-will, an American nightjar. It sleeps for much of the winter, but its temperature does not drop far.

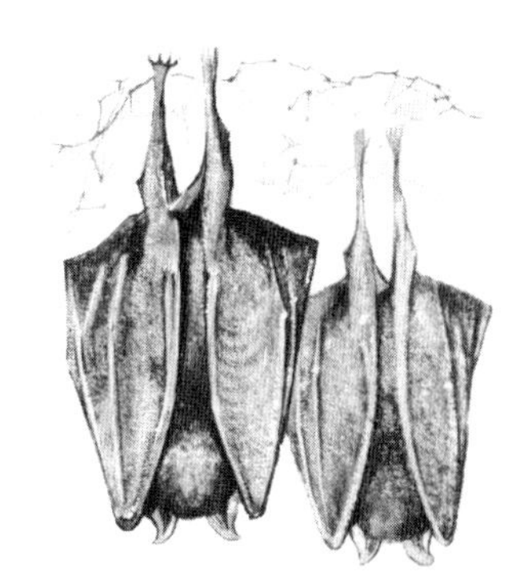

Right: The bats of cool climates sleep throughout the winter in caves and other secluded places. They hang upside down and their temperature falls to a very low level. Moisture from the air often condenses on their fur and gives them a glistening appearance.

Many mammals sleep for several days at a time during the winter but this is not true hibernation. True hibernation, where the body temperature falls almost to that of the surroundings, is found in only a few groups of mammals. Bats and other insect-eating mammals such as hedgehogs hibernate, and so do various groups of rodents. These include dormice, ground squirrels, and hamsters.

Colder weather, shortage of food, and the shorter days of autumn probably play a part in bringing on the winter sleep. In spring the animals gradually speed up their body processes until their temperatures return to normal and they can move about. The waking process uses up a great deal of energy, and it is likely that many hibernating animals never wake up because they have not enough food reserves to last the winter.

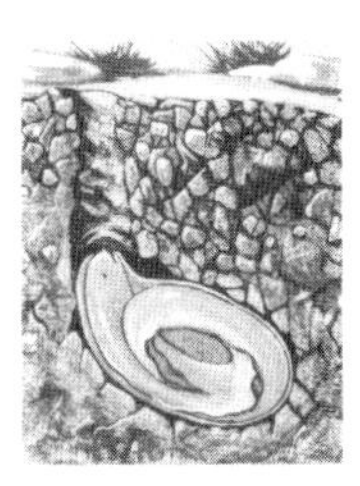

The lungfish aestivates in the mud of its dried-up river.

Aestivation—the Summer Sleep

In warmer parts of the world, especially where there is a dry season, a number of animals go to sleep to avoid the heat and drought. This is called *aestivation*. The African lungfish lives in sluggish rivers which often dry up. The fish burrows into the mud and surrounds intself with mucus. It breathes air through a narrow 'chimney' and it remains moist until the rains come.

The Canada goose migrates between Alaska and Mexico.

The white stork, a summer visitor to Europe.

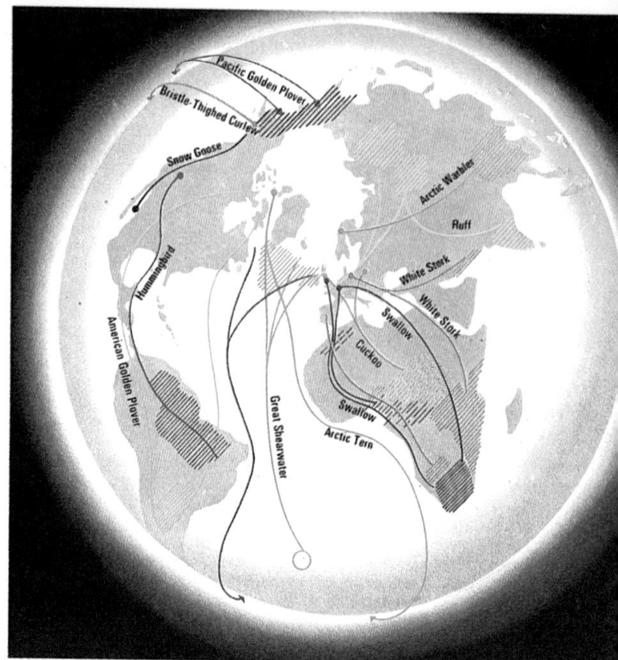

The migration routes of many birds have been discovered by putting little rings on their legs and recording where they have been found. As the map shows, some species travel huge distances.

Migration

The cuckoo is one of the many birds that spend the winter in the tropical and sub-tropical regions and return to temperate regions in spring to breed. This regular movement between two regions is called migration. It is of great importance to the birds because it enables them to find food throughout the year. Nearly all these migrants are insect-eaters whose food disappears in winter. Other birds, such as ducks and waders, migrate between temperate and polar regions, especially the Arctic.

How migrants find their way is still something of a mystery. They probably navigate by a combination of the sun and stars, the Earth's magnetic field and landmarks.

Other Migrants

Land-living mammals are sometimes migratory. The caribou, for example, spends the summer in the Arctic tundra, and moves to the conifer woods farther south for the winter.

The purple martin travels between Canada and Brazil.

Many butterflies migrate. The American monarch spreads far into Canada, then in early autumn the butterflies fly southward in huge swarms to Florida and Mexico.

Migrants in the Sea

Many whales move from their polar feeding grounds to give birth in warmer seas. The European eel starts its life in the Sargasso Sea of the Western Atlantic. It then takes three years to reach the coasts of Europe and swim upriver. Several years later, the fully grown eel swims back to the Sargasso to breed and die. Salmon do the opposite. They swim upriver to breed, after feeding at sea.

Some animals migrate every day. Crustaceans and other members of the marine plankton spend about 12 hours in the surface layers and then sink to deeper water for the rest of the day.

Nocturnal Life

Above: The fox hunts mainly by night, using its keen eyes, its ears, and its sense of smell.

At night the temperature drops and the air becomes damp. This is very important for small creatures such as slugs, snails, and woodlice which have porous coats. They spend the day hidden away but can come out at night without risk of drying up. Most desert-dwelling animals also come out at night when it is cooler and get much of their water by eating dew-laden vegetation. Nocturnal life is also a way of avoiding predators which hunt by day.

Many nocturnal animals have large eyes, which gather as much light as possible. The owl is a good example, but our own eyes are also good for seeing at night, although the cells which detect colour do not work in dim light and we see everything in shades of grey. Cats, dogs, and some other animals increase the sensitivity of their eyes with a reflective tapetum. Light is reflected back through the light-sensitive retina so that it is stimulated twice and produces a brighter image.

Left: The bat has weak eyes and it finds its way about by sending out high-pitched sounds and listening for the echoes. The directions of the echoes tell the bat whether anything is in the way or not.

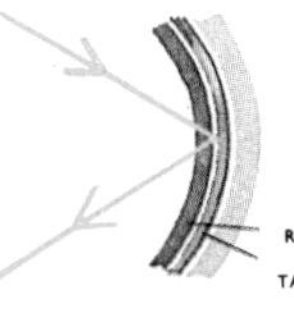

Cats have a 'mirror' behind the eye and this makes the eye much more sensitive. During the daytime the pupil is reduced to a narrow slit, preventing too much light from entering.

Not all nocturnal animals have good eyesight, however. Many rely on other senses. The sense of touch may be developed in the form of long whiskers, as in cats and mice. The sense of smell is also very acute in badgers and hedgehogs. Hearing is used for detecting prey and enemies, and sounds are used for communication. Many of the sounds are beyond the range of human hearing.

The Bat's Sonar

The insect-eating bats fly by night. They have a system of sonar or echo-location, which is rather similar to radar. The bats give out a stream of very high pitched sounds—far above our hearing—and they listen for the echoes bouncing back from nearby objects. The echoes are picked up by the bats' ears and used to pinpoint obstacles and insect prey.

Animal Lights

Many animals produce their own light, mostly to attract mates. The European glow-worm is a beetle whose female has a light-producing organ on her abdomen. The greenish light is produced by a chemical reaction. Man-made lights give off heat but the glow-worm's light-producing reaction is very efficient and gives out almost no heat. Fireflies are also beetles. Both sexes produce light, and exchange flashing lights in their courtship.

Light-producing organs are common in marine animals, especially in the darkness of the depths. Jellyfishes, prawns, squid, and fishes have luminous species. The lights help them to find mates and food. Deep-sea angler fishes have luminous 'bait' on 'rods' placed on their heads. They lure smaller fishes to their mouths. Some squids and prawns give off luminous slime to baffle their enemies.

Animal Electricity

Certain fishes have the remarkable ability to generate high voltage electrical energy in their bodies, sufficient in some instances to stun a horse. The fishes use these electric shocks for paralysing their prey and for defending themselves. They can also use them for navigational purposes.

All animals produce electrical energy, for an electric impulse is involved every time a signal travels along a nerve and into a muscle. The electric fishes have simply modified this system.

A normal muscle contracts when it receives an electrical signal from a nerve ending. The muscles of the electric organs cannot contract, however, and the energy is not

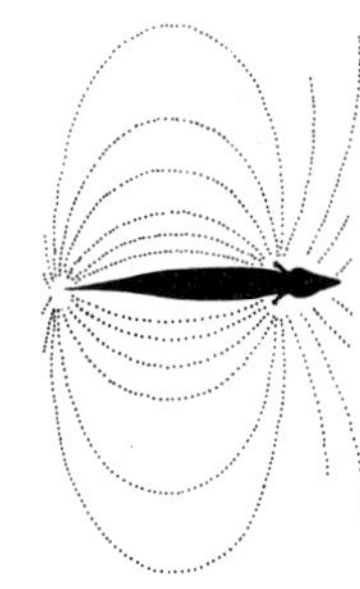

Above: The pattern of the electric field surrounding the Nile fish. The pattern is disturbed when the fish approaches an object, so the fish avoids it.

Below: Some well known electric fishes.

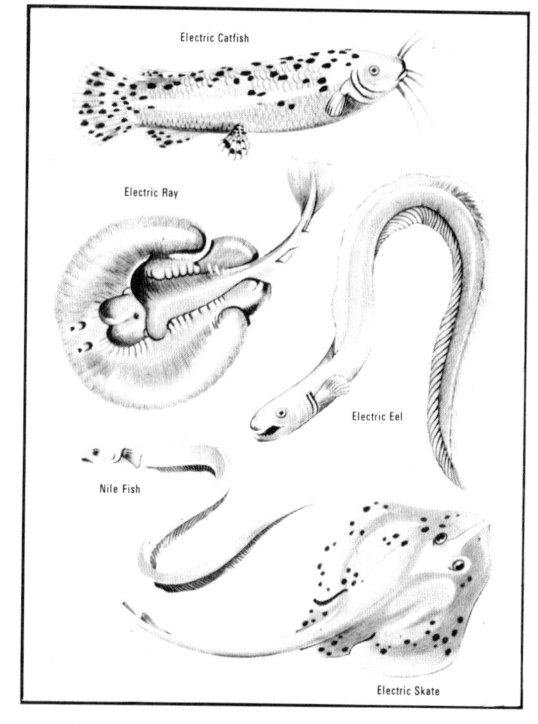

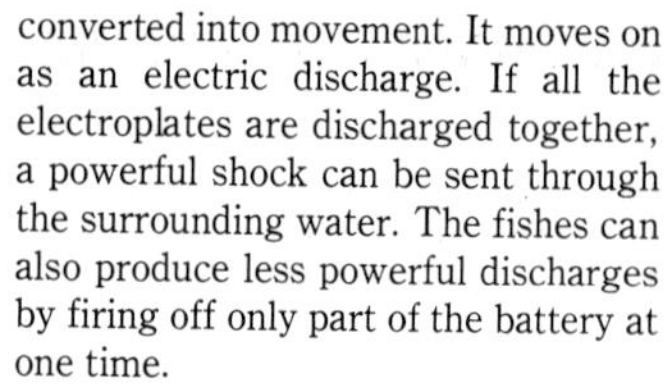

converted into movement. It moves on as an electric discharge. If all the electroplates are discharged together, a powerful shock can be sent through the surrounding water. The fishes can also produce less powerful discharges by firing off only part of the battery at one time.

Electric organs are found in a variety of fishes, in the sea and in fresh water. One of the best known of the electric fishes is the electric eel, which lives in the rivers of the Amazon basin in South America. The electric organs run about three-quarters of the way along the body and they can generate up to 550 volts—more than enough to kill frogs and fishes. Larger animals coming too close to the eel are likely to be stunned by the shock.

The eel-like Nile fish also finds its way about in muddy water by means of its own electricity. It surrounds itself with an electric field generated in its tail. It is very sensitive to disturbances in this field so, if it swims too near a rock, it will detect a change in the field and take avoiding action. The fish can also use this system to seek out crevices for safe hiding.

The electric organs of the electric catfish form a sheath around most of the body. They generate about 350 volts and are used for stunning the catfish's prey. The electric catfish lives in the rivers of tropical Africa.

Marine electric fishes normally generate lower voltages than fresh-water fishes. This is because salt water is a much better conductor of electricity than fresh water. It requires a lower voltage to push a current through it. The skates and rays are the most important electric fishes in the sea. Most of the species have batteries in their tails and produce low voltages of unknown function. The various species of electric rays, however, have large batteries on each side of the head. They produce a large current at up to 200 volts. The discharge is used for both defence and feeding.

Skeletons and Muscles

Most animals have a skeleton. Our own skeletons, like those of all other back-boned animals, are inside our bodies. But some animals have skeletons on the outside of their bodies. These animals include insects, crabs, and snails.

Are Skeletons Necessary?

The skeleton takes the weight of the body and the other parts of the body are normally suspended from it. The skeleton also gives shape to the body. This is especially obvious in crabs and other animals with skeletons outside.

The skeleton also enables animals to move because it is an anchor for the muscles. The other main job of the skeleton is to protect the animal. Our skulls protect our brains, and our ribs protect our hearts and lungs.

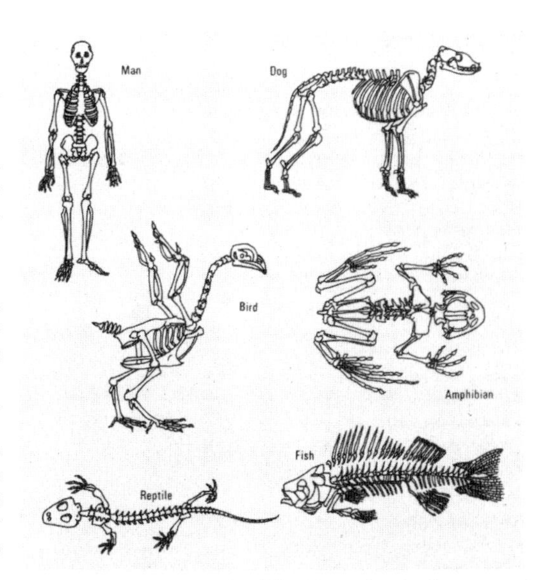

Vertebrate skeletons all have the same basic pattern, but it is adapted for different ways of life.

Joints

The skeleton is made up of numerous sections which meet at *joints*. When two bones meet at a joint they are held together by very tough straps called *ligaments*. The ends of bones are covered with smooth cartilage and they are often enclosed in a bag of liquid to help the bones to move smoothly against each other.

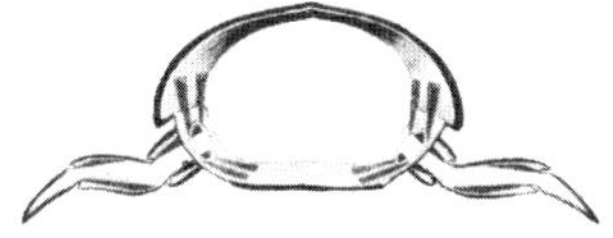

The crab does not have a backbone: its skeleton consists of hard plates on the outside of its body.

Muscles

Muscles consist of thousands of tapering fibres collected into bundles. Most muscles are connected to the skeleton by each end. Vertebrate muscles are attached to bones by means of very strong cords called tendons.

When a nerve stimulates a muscle the fibres contract rapidly. Then the muscle relaxes and can be stretched again, but it cannot expand and push out. Muscles therefore have to work in pairs. The biceps bends the arm and the triceps straightens it again.

Insect wings are made to vibrate very rapidly by the contraction of two sets of muscles in the thorax.

Muscles are not always very close to the bones which they pull. The bones of a bird's leg and foot are worked by muscles high up in the leg. These muscles are joined to the foot bones by long tendons—the sinews that you remove when preparing a chicken for the oven. The lower part of the leg of a horse contains very little muscle. It is worked by large muscles at the top of the leg. The lower part is quite light so less effort is needed to move it.

The muscles that help us move, the *skeletal* muscles, also called *voluntary* muscles because we can order them to work, can contract quickly. There are two types: red and white. The meat on a chicken's breast is white muscle, but the dark flesh of its legs is mainly red. Red fibres need plenty of oxygen and are good for sustained work. White fibres are good for violent activity but they tire quickly. *Involuntary* muscles continue working without our knowing. They include the muscles which push food along the intestine. They contract slowly and use little energy.

The Senses of Animals

Animals use their eyes, ears, and other sense organs to find out what is going on in their surroundings. Signals are sent from the sense organs to the brain and the brain then instructs the body to react accordingly.

The simplest animals, such as *Amoeba*, have no special sense organs. The whole body surface of these creatures is sensitive and they react to unpleasant sensations by moving away. The more advanced animals have a number of special sense organs, each constructed for picking up different kinds of signals from the surroundings. There are also nerves, which carry signals from one part of the body to another, and a brain which is the control centre. Signals are received in the brain and the brain then sends signals to the body telling it to take the correct action.

We tend to think that we have only five senses: sight, hearing, taste, smell, and touch, but there are several more. The senses of balance and temperature, for example, are quite important. And there are senses that tell us about the conditions inside our bodies. The sense of pain tells us when something is wrong, and other nerves in our bodies tell us when we are hungry or thirsty. Similar kinds of sense organs are found in other animals, although they are not always made up in the same way and they are not always on the same part of the body. The senses are not equally developed, and some of the sense organs may be absent.

The Eyes

The eyes of vertebrate animals are all more or less alike. The basic parts are the *lens* and the *retina*. The lens gathers the light rays and focuses them on to the retina. The latter consists of millions of light-sensitive cells. When light falls upon them, they send signals to the brain. The brain then interprets the signals as a picture.

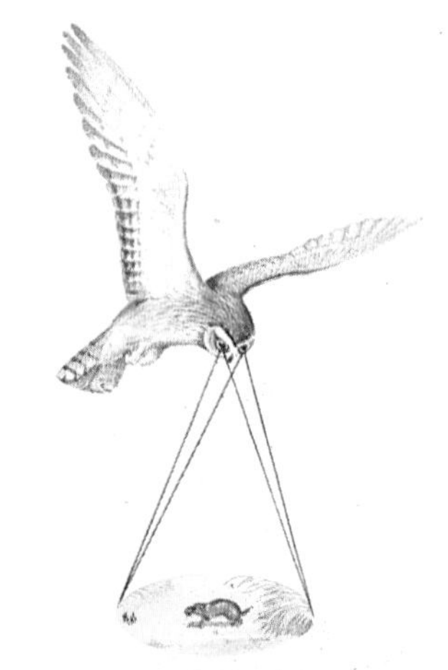

Hunting animals need to be able to judge distances well. Two eyes at the front help them to do this. The owl's large eyes help it to see well at night.

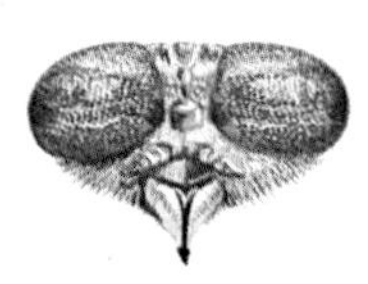

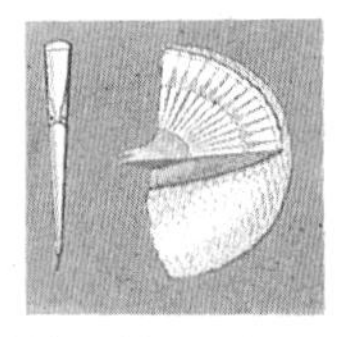

Insect eyes are made up of lots of small cone-shaped lenses.

Insect eyes are constructed on a very different pattern to the vertebrate eye. They are known as *compound eyes* because each one is made up of numerous tiny units. Each unit is cone-shaped and they fit together so that their lenses form a patchwork pattern on the surface of the eye. Each lens sends its own signal to the brain, so that the insect sees a rather blurred mosaic image. Although the image is not very clear, the insect easily spots movements.

The Ears

Sounds play a large part in the lives of many animals. They are used as warnings and as signals to attract

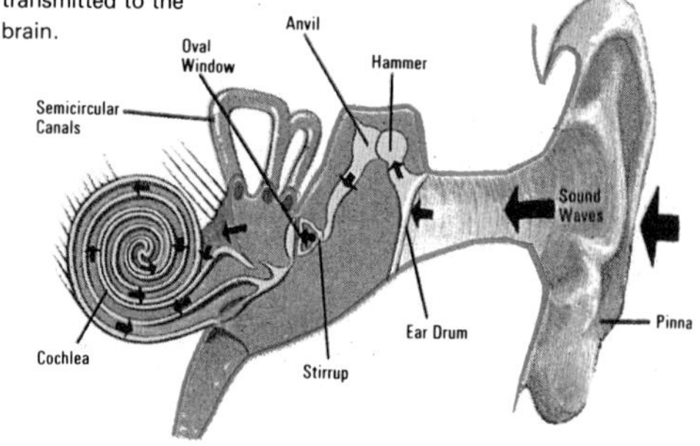

Sounds reaching the human ear drum set up vibrations in a chain of tiny bones. These carry the sounds to the inner ear, where the vibrations are picked up by nerves and transmitted to the brain.

mates. The sound-receiving organs, or ears, are normally membranes stretched like drumskins across an opening. They vibrate when sound waves hit them and send signals to the brain. Some insects pick up sounds with sensitive hairs, and crickets have ears on their front knees.

The impala must always be alert to danger. Even while drinking, its eyes are scanning the area and its ears are pricked up waiting for the faintest sound of danger.

Chemical Senses

The senses of taste and smell are chemical senses. The sense of smell tells us about chemicals in the air. Most substances give off vapour or scent and this consists of minute particles which affect our smelling or olfactory organs. Insects smell with their antennae, and some male moths have such sensitive smell receptors that they can smell female moths over 1 kilometre away.

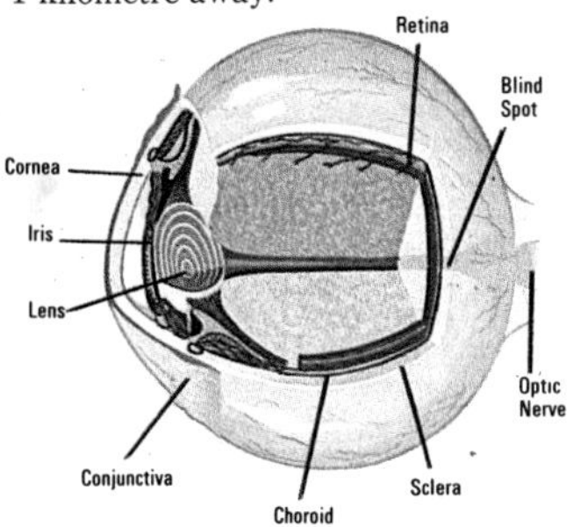

Right: The arrangement of the nervous system is very similar in man and the frog, although man's brain is very much larger and more efficient.

Left: Man's eyes are his most important sense organs. Light is gathered by the lens and focused on to the retina at the back. Nerve cells there are stimulated and they send messages to the brain.

The sense of taste tells about chemicals in contact. We can taste things only with our tongues, but flies, for example, have taste organs on their feet.

The Sense of Touch

Hairs normally have tiny nerves wrapped around their bases and increase the sense of touch. Many animals use hairs as special organs of touch. Mice and cats have very sensitive whiskers which help them to find their way in the dark.

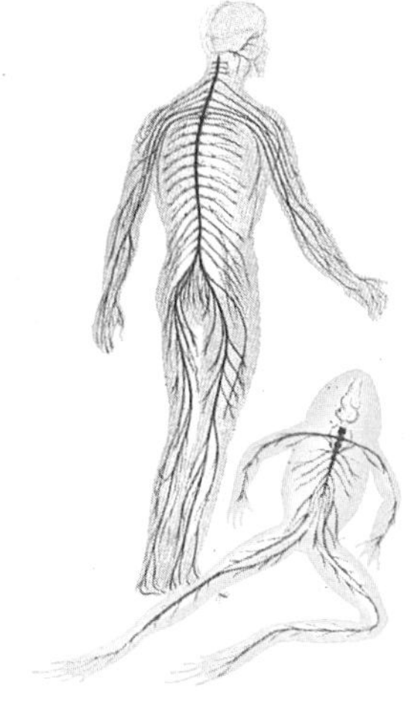

Feeding and Digestion

The waterbuck eats grass and other leaves and is a typical herbivore.

The food provides energy and the materials for building up the body. Animals rely on green plants to provide their energy. They eat plants or other animals, but the energy always comes from a plant to start with.

Some animals feed entirely on plants. They are called herbivores. Some species feed entirely on other animals. They are called carnivores or flesh-eaters. Other species feed on both plants and animals. They are called omnivores.

Herbivores

Herbivorous or plant-eating animals include voles, rabbits, deer, sheep, horses, snails, caterpillars, and many

The toad is a carnivore, feeding on slugs and insects which it catches with its long sticky tongue.

others. Most feed on the leaves and other juicy parts of the plants, but some animals feed on wood. Examples include the termites and many beetles.

Carnivores

Carnivorous or flesh-eating animals include cats, dogs, weasels, owls, snakes, frogs, sharks, ladybirds, starfishes, sea anemones, and many other creatures. Some animals chase their prey, others lie in wait until something comes along. Some spiders and some caddis fly larvae set traps for their prey by spinning silken webs or snares. Snakes and other animals use poison to kill or paralyse their food.

Most carnivores catch their food while it is alive, but some feed on dead animals, or carrion. Carrion feeders include vultures and hyaenas. Many insects also feed on dead animals, including the burying beetles or sexton beetles. They bury dead animals by dragging soil out from under them and then lay their eggs in them.

Omnivores

Many herbivorous animals will eat flesh sometimes. For example, voles and other rodents will often eat insects. Carnivores may also eat vegetable food from time to time. But there are omnivorous animals which regularly eat a mixture of plant and animal materials. They include bears, badgers, rats, and cockroaches.

Filter Feeders

Many animals living in water feed on tiny particles which they filter out from the water. These particles might be tiny plants or animals, or else they might be decaying matter falling to the bottom of the sea or the pond. The filter feeders are therefore omnivorous animals. Bivalve molluscs have two tubes or siphons, one to suck in water and the other to squirt it out. The water is strained or filtered and the particles held back are passed to the mouth. The animals obtain oxygen from the water current at the same time as they strain out their food. Freshwater bivalves are useful in fish tanks because they filter out many of the microscopic algae which would otherwise turn the water green.

Some marine worms are also filter feeders. They have crowns of feathery arms or tentacles around their heads and these strain out particles falling from above.

Animal Teeth

The teeth of fishes, amphibians, and reptiles are simple spikes. Birds have no teeth at all today, although some extinct birds had them. The most complex teeth are found in mammals. Their job is to catch, cut, and chew the food. The number and arrangement of a mammal's teeth depend upon its food. Flesh-eaters, such as dogs, have large eye teeth (canines) near the front. These stab the prey. The back teeth have sharp edges and slice the meat into pieces. Grazing animals have back teeth with broad flat surfaces suitable for grinding up the grass.

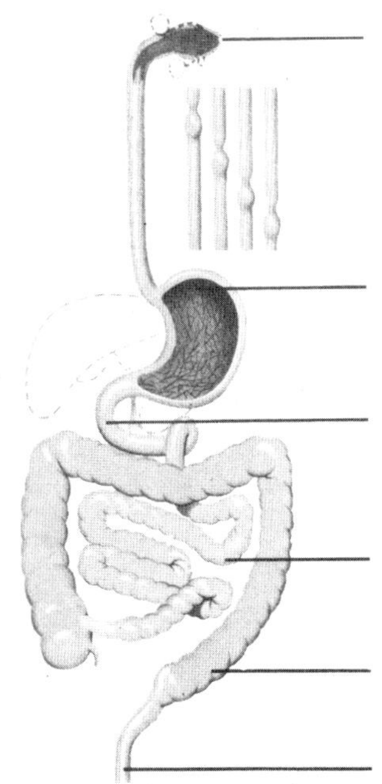

Digestion begins in the mouth, where the salivary glands pour saliva on to the food.

In the stomach the food is churned up with dilute acid and digestive juices. Food particles are reduced in size and the chemical breakdown of protein begins.

In the duodenum the food meets other digestive juices, including some from the liver and pancreas.

Chemical breakdown is completed in the small intestine and the food materials are taken into the blood.

Undigested material passes into the large intestine or colon and water is reabsorbed.

The remaining material passes into the rectum.

Digestion

When an animal has eaten its food it must convert it to some usable form. This conversion is carried out in the digestive system. The food is broken down into simple substances which can be absorbed and carried around the body in the blood, then used to provide energy or to build new cells. The detailed structure of the digestive system varies a great deal, but it is basically a tube which begins at the mouth and ends at the anus. Food is taken in through the mouth, and the front part of the system is concerned mainly with storing and breaking down the food into smaller pieces and simpler substances. The next part of the system is concerned with absorbing the digested food into the blood, and the last part is concerned with getting rid of the undigested material or waste. Sea anemones, jellyfishes, and some other simple creatures have only a mouth and this must serve for taking in food and passing out waste. Some parasites absorb food through the skin.

GETTING RID OF WASTE

Waste products must be removed by excretion. The main organs involved in excretion are the kidneys. Each kidney consists of about a million little tubes. Blood coming into the kidney is filtered through these little tubes. Unwanted materials, including excess water, flow into the kidney tubes and into a large collecting duct called a ureter. This carries the waste materials (now called urine) from the kidney to the bladder.

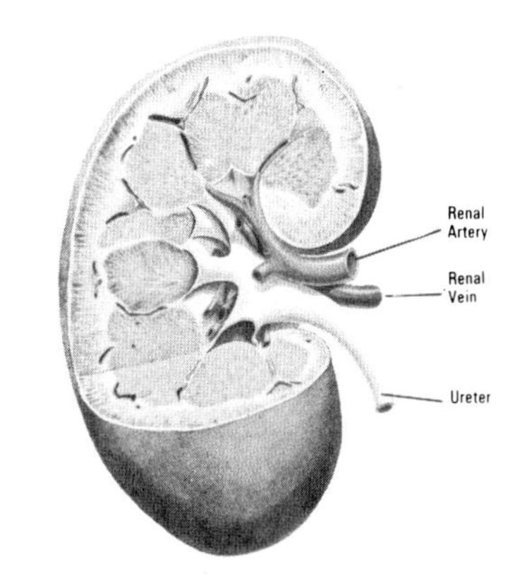

How Animals Breathe

Breathing involves taking in oxygen, which is then used to 'burn' the food in the cells of the body and to release energy. The whole process of obtaining oxygen and using it in the body is called *respiration.* One of the waste products of respiration is a gas called carbon dioxide. The breathing organs get rid of carbon dioxide as well as obtaining oxygen for the animals.

The simplest animals, such as the protozoans and the jellyfishes, have no special breathing organs. Larger and more active animals need special breathing organs and they also need a transport system to carry the oxygen around the body. The blood stream is the transport system.

Breathing under Water

Many aquatic animals come to the surface to breathe air. Examples include whales, seals, and some water snails. Many others, however, remain under the surface and rely on oxygen dissolved in the water. The larger ones have special breathing organs called *gills.* In common with all breathing organs, the gills have a large surface area through which to absorb oxygen.

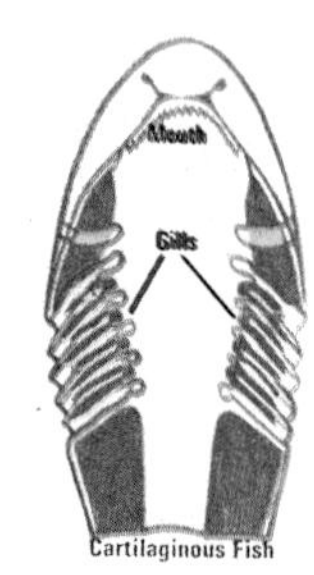

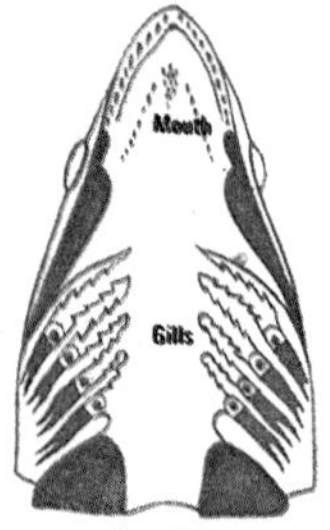

Bony Fish

The fish's gills lie in pouches at the back of the throat. In the shark-like fishes each gill has its own separate opening on the side of the body. The gills of the bony fishes are covered by the operculum.

Fish gills consist of columns of finger-like organs situated in cavities connected to the throat. The cavities also open through a number of slits on to the sides of the body. The fish takes in a mouthful of water and then, by shutting its mouth and raising the floor of its throat, it forces the water over the gills. The blood arriving at the gills has relatively little oxygen in it, and so oxygen tends to pass into the blood from the water.

Other animals breathing by means of gills include the crabs and lobsters, the bivalves, and snails. The gills of the crabs and lobsters are rather feathery outgrowths covered by the shell. One of the limbs wafts a continuous current of water through the gill chamber.

The gills of bivalves are very large and they serve to trap food particles as well as to absorb oxygen. The animals have two breathing tubes or siphons. One brings a current of water into the gill chamber and the other carries the water away. Many water snails, including most of those that live in the sea, breathe by means of gills. Water is pumped in and out of the gill cavity and the gills absorb oxygen from it. Land snails have lost their gills, but they still have the gill cavity. It is lined with blood vessels and acts as a lung with the snails pumping air in and out of it. Some snails return to the water after a period of life on land. Most of them still breathe air and they can be seen coming to the surface now and then to renew the air in their lungs.

Birds are very active animals and they need a good supply of oxygen. Their ribs and keel move backwards and fowards, forcing air in and out of the lungs and air-sacs.

Air-breathing Vertebrates

The reptiles, birds, and mammals breathe by means of *lungs.* These are thin-walled sacs, rather like balloons, and they lie in the chest cavity. Air is pumped in and out of them and oxygen is absorbed. Most amphibians also have lungs, although these are rather inefficient organs. The amphibians are moist-skinned animals and they can absorb a good deal of oxygen through the skin. The lining of the frog's mouth and throat is also very thin and well supplied with blood vessels. Much oxygen is absorbed in this region.

Reptile lungs are much more efficient than those of amphibians and the reptiles cannot breathe through their skin. The lungs are elaborately folded and they have a fairly large surface area. This folding is taken even further in the lungs of mammals. The lung is made up of thousands of tiny air pockets, each of which is surrounded by blood vessels. This arrangement provides a very large surface area for the absorption of oxygen, but it does mean that there is a good deal of stale air in the remote parts of the lung. Not all the air can be exchanged at each breath.

Birds are extremely active animals and they need large amounts of oxygen. Their lungs are extremely efficient. Instead of being closed at the end, the lungs lead to thin-walled air-sacs. When the bird breathes in, the air passes through the lungs and into the air-sacs. At each breath, air in the lungs is completely changed and there is never any stale air.

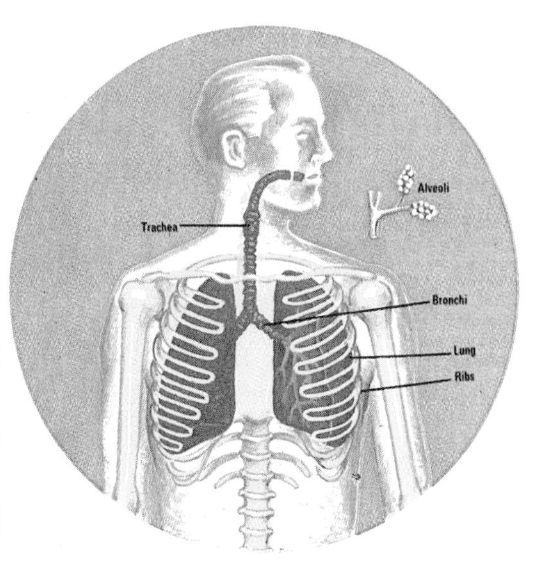

The human lungs shown in position in the chest cavity. The small diagram shows how the lung tubes end in clusters of thin-walled pouches where oxygen is absorbed.

Insect Breathing Systems

The insects, together with the centipedes and a few other arthropods, have a breathing system quite different from that of other animals. The body is penetrated by a maze of tiny, air-filled tubes called *tracheae*, which open on to the body surface at holes called *spiracles*. The inner ends of the tubes normally contain liquid, but this is reabsorbed when the insect is very active. The air can then get further along the tubes and provide the extra oxygen needed for the activity.

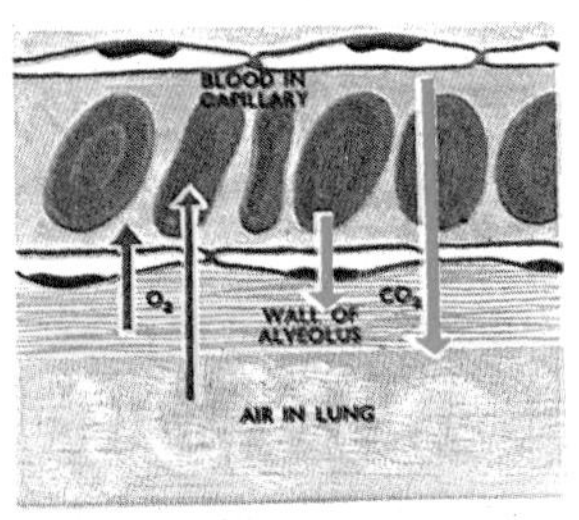

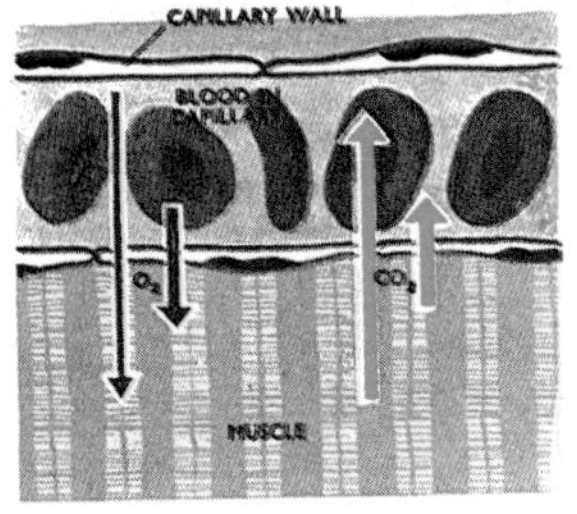

Left: Oxygen concentration in the lungs is higher than in the blood. Consequently oxygen spreads, or *diffuses*, through the walls of the alveoli into the blood. There is more carbon dioxide in the air than in the lungs, so carbon dioxide passes from the blood into the lungs. At the tissues the situation is reversed.

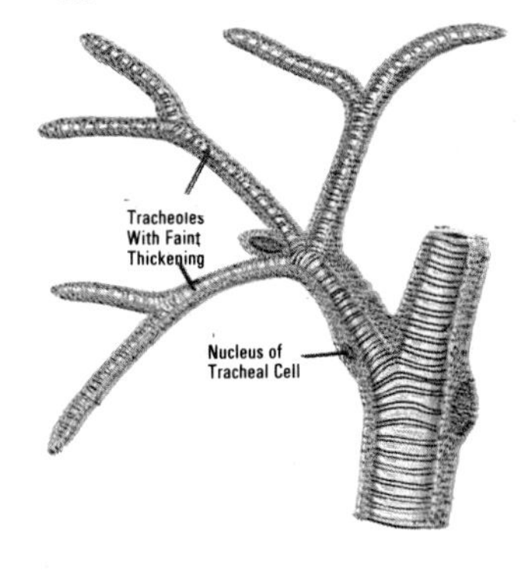

Right: Part of an insect's breathing system. The fine tubes carry oxygen to all parts of the body.

The Blood System

All but the simplest animals have some sort of blood system. It is the body's transport system, and carries food from the digestive organs to all parts of the body. It also carries waste products from the tissues to the excretory organs. In most animals the blood carries oxygen from the breathing organs to the tissues. Hormones—chemical 'messengers'—are also distributed by the blood. A less obvious function is the distribution of heat. Yet another function of the blood is fighting infections and destroying germs.

Mammalian Blood

Mammalian blood, like that of the other vertebrates, consists mainly of a liquid called plasma. Floating in the plasma there are millions of tiny cells called corpuscles. The plasma is a pale yellowish colour. It is mainly water, but it contains a very complicated mixture of proteins, sugars, and dissolved minerals.

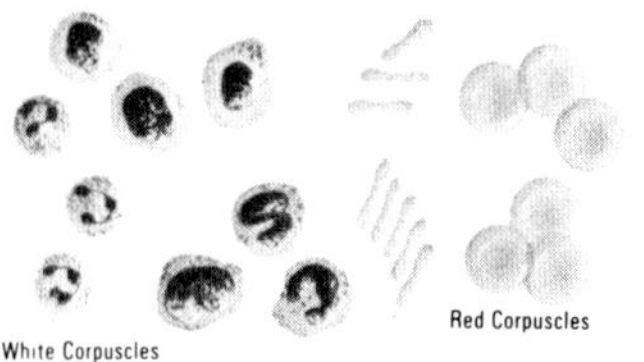

Left: Red and white blood corpuscles. The red ones have no nuclei.

The blood corpuscles are of two main types: red and white. Red corpuscles, which give the blood its red colour, are minute disc-shaped cells. Their job is to carry the oxygen around the body. The red colouring matter, haemoglobin, takes up the oxygen in the lungs and releases it again as the blood flows around the body (see page 98). Mammalian red corpuscles are made in the bone marrow and have no nuclei. They do not live very long—about 100 days in man—and new ones are always being made.

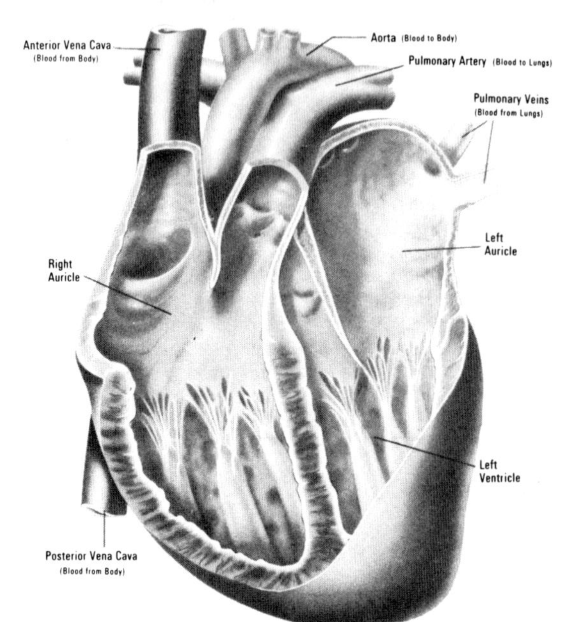

Above: A drawing of the human heart showing the various chambers and the valves which ensure that the blood flows in the right direction.

White blood corpuscles are concerned with the control of infection. In other words, they fight germs that get into the blood. Some of the white cells are rather like amoebae (see page 50) and they swallow up the germs. Other white cells produce *antibodies*. These are proteins which kill the germs. White cells also help to heal wounds.

In addition to the red and white corpuscles, the blood contains some smaller bodies called *platelets*. These break down when they come in contact with the air and they cause the blood to clot and help to form scabs.

The Heart and the Blood Vessels

The insects and other arthropods have an open blood system. The blood lies in large cavities and surrounds the internal organs. Most animals, however, have a closed blood system with the blood running in vessels.

The blood has to circulate round the body, and so the blood system has a pump, the heart. The shape and structure of the heart vary from one animal group to another, and it is divided into two or more chambers. The blood vessels that carry blood away from the heart are called *arteries*. The main arteries have thick, muscular walls. When they reach their destination they divide up into tiny vessels called *capillaries*. These have very thin walls and materials can pass very easily between the blood and the tissues. The capillaries then join up again to form the *veins* which carry the blood back to the heart. The veins have thinner and less muscular walls than the arteries and the blood pressure is much lower.

The diagrams below show the arrangement of the major blood vessels in various vertebrate animals.

Warm-blooded or Cold-blooded?

The birds and mammals keep their bodies at constant high temperatures and are called warm-blooded. Fishes, frogs, snakes, and the invertebrates are unable to keep their temperatures up and are called cold-blooded. But they are not always cold. Their bodies are more or less at the temperature of their surroundings. A snake basking in the sunshine will actually be quite warm. Birds and mammals use up a lot of energy to keep their temperatures up. Cold-blooded creatures need less energy and they can go without food for long periods.

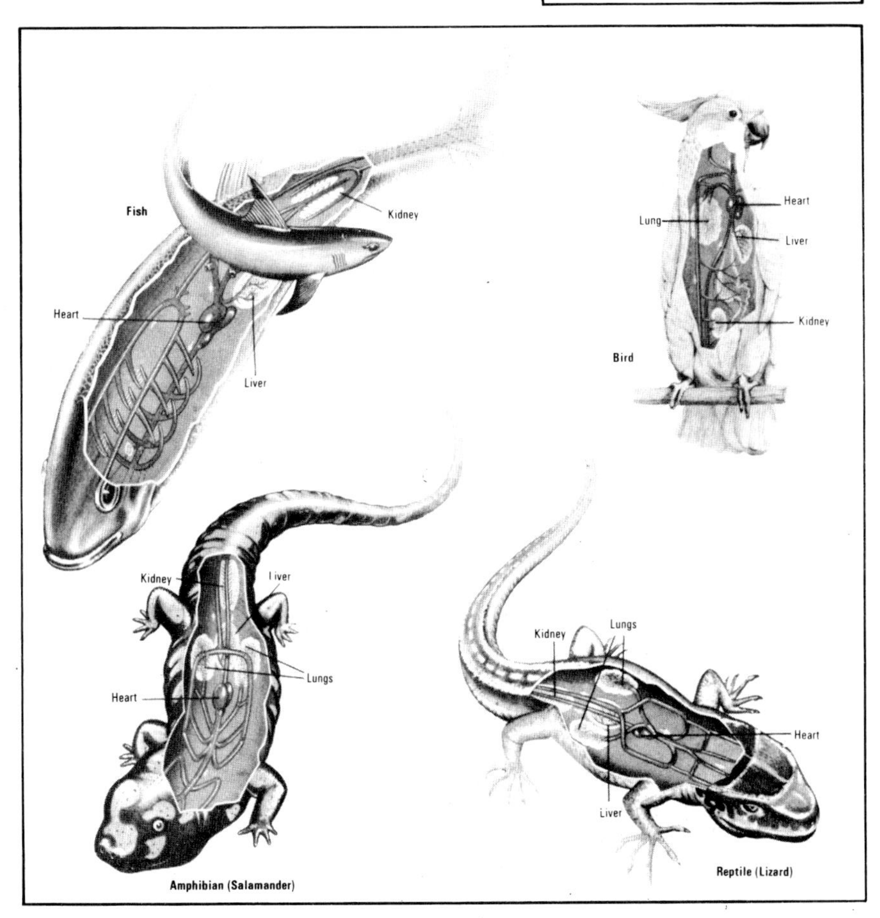

Reproduction

Reproduction is one of the characteristics of living things. There are two main methods of reproduction—sexual and asexual. Sexual reproduction, which is the more important of the two, involves the joining together of special cells called gametes or sex cells. Asexual reproduction takes place without any joining of cells and often involves little more than the breaking of an animal into two parts. Almost all animal species reproduce sexually at some time or other, but only simple animals reproduce asexually.

Splitting in Two

The simplest form of reproduction takes place in some of the protozoans, such as *Amoeba*. The cell goes on growing until it reaches a certain size—perhaps until it gets too big for its nucleus—and then it divides into two. In a sense, these creatures have eternal life because they do not die from old age, but large numbers are eaten by other creatures.

Budding

Some simple animals, notably the corals and their fresh-water relative *Hydra*, produce new individuals by budding. A 'bud' develops somewhere on the body and gradually grows to form a branch.

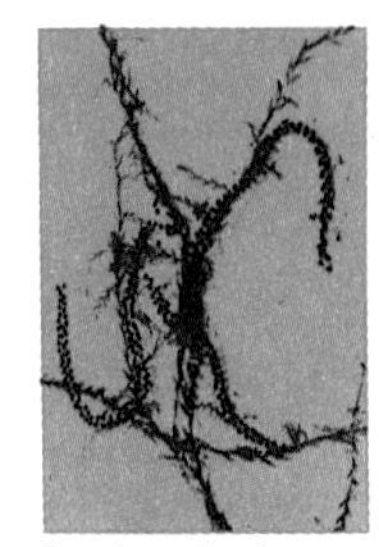

Female toads take no interest in their offspring. They lay large numbers of eggs, but only about two will ever reach maturity.

Earthworms contain both male and female organs, but they have to pair up before they can lay eggs. Some species pair on the surface of the ground at night.

Below: *Hydra* has two methods of reproduction. New individuals can grow out as buds (upper pictures) or the animal can reproduce sexually by producing eggs and sperm.

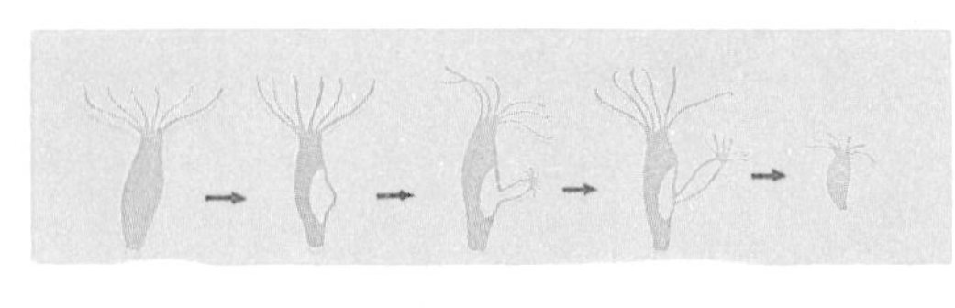

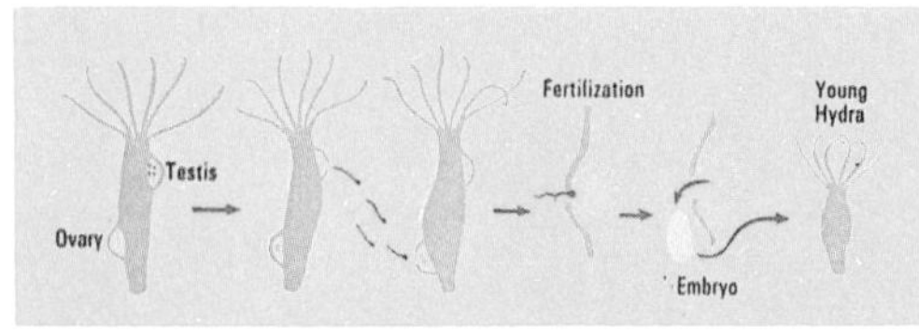

Sexual Reproduction

This method of reproduction involves the joining together of gametes. The joining together of the gametes is called fertilization. In *Amoeba* and some of the other simple animals all the gametes are alike, but the more advanced animals have two kinds of gametes. One is called a sperm and the other is called an egg or ovum. Sperms are produced by the male sex and eggs are produced by the female sex.

Some water-dwelling animals merely release their sperms and eggs into the water and leave fertilization to chance. In most animals, however, the two sexes have to meet before reproduction can take place. Some animals give out a scent which attracts the opposite sex. Others rely on elaborate displays to bring the sexes together

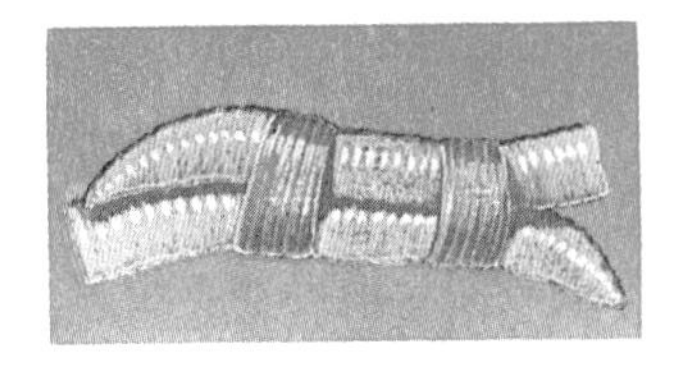

and to ensure that eggs and sperms are produced at the right time. The eggs may be fertilized inside the female's body or outside it.

When an egg has been fertilized it starts to develop into the new animal. The egg contains enough food material for all the early stages of development. Eggs that have been fertilized inside the mother's body may be laid soon afterwards, or they may remain in the mother's body. Quite often they hatch before they leave the mother, in which case the mother gives birth to active young. In most of the mammals, the group of animals to which we ourselves belong, the young animal remains inside its mother long after development from the egg started. It obtains food from the mother's own body through the umbilical cord.

Parental Care

Some animals produce large numbers of eggs or young, while others may have only one or two babies. The number of eggs or young depends very much on the dangers that they face during their lives. Animals that merely scatter their eggs and abandon them suffer the most, and these animals survive only by producing huge numbers of eggs. The female cod, for example, may produce about 9 million eggs, but only two can be expected to reach maturity. Animals that look after their offspring can get by with far fewer young.

Hermaphrodites

The earthworms and many snails do not have separate sexes. Each individual possesses both male and female reproductive organs. Such animals are called hermaphrodites but they have to pair. Each then goes away and lays its own eggs, fertilized by sperms from the other animal. This arrangement is very useful for slow-moving creatures such as worms and snails. They can pair with any adult of the same kind. The chances of finding a mate are therefore greatly improved.

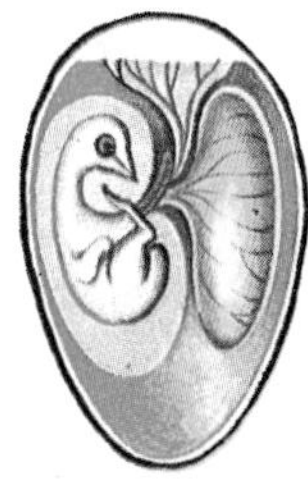

A bird's egg contains just the right amount of food for the development of the young bird. The bird completely fills the egg when it has used up all the food.

Emperor penguins breed in large colonies on the frozen Antarctic sea and carry their eggs and chicks on their feet.

Parthenogenesis

Many stick insects, aphids, and other insects can lay eggs which hatch without ever being fertilized. The male sex is very rare or even unknown in some species. This is called parthenogenesis. We can think of it as a special kind of sexual reproduction because it involves the production of eggs.

Regeneration

A number of the simpler animals are able to regrow or regenerate parts of their bodies that have been lost or damaged. In some animals this ability amounts to a form of asexual reproduction. *Hydra*, for example, will not normally die if it is cut into two parts. Each part will regenerate the missing sections and become a complete animal. Flatworms are also very good at regenerating themselves. They can be cut into several small pieces and each piece will normally grow into a new animal. Starfish can grow new arms if any are lost, and the broken arms can sometimes grow new bodies!

Animals with backbones can repair cuts and wounds, but they cannot regenerate lost limbs. Lizards can break off their tails if they are alarmed. A new tail will grow, but it is normally shorter than the old one.

Homes and Habitats

Each species has its own preferred *habitat,* another way of saying surroundings. *Environment* is another word meaning much the same thing. The climate and the soil are the most important factors because they control the vegetation. The vegetation in turn affects the distribution of the animals. The study of habitats and of the plants and animals that live in them is called *ecology.*

Some of the Earth's major habitats are described in the following pages, together with examples of how the plants and animals are adapted for life under various conditions.

The Earth provides a variety of homes for living things ranging from the cold tundra (above), through forest retreats (left) and open grasslands (right) to deserts (below). In all types of environment plants and animals are found adapted to the prevailing conditions.

Life in the Rain Forests

Rain forests develop where the temperature is always above 18°C and the rainfall is at least 200 centimetres per year.

Hundreds of Species

Rain forest plants grow throughout the year and all are evergreen. The abundant moisture means that they grow quickly. There may be as many as 300 species of trees in one square kilometre of forest. Animal life is also very abundant and we may never know how many species there are.

All rain forests have the same structure. The roof, or *canopy,* of the forest is about 30 metres high, with scattered taller trees, and a lower layer at 15 metres.

Along river banks and where the forest has been felled, the vegetation at ground level is very thick. This is the 'jungle', but farther into the forest, the undergrowth disappears because there is so little light.

Treetop Plants

Most of the rain forest's life is among the branches of the canopy. The leaves and flowers of the trees are here, and there are thousands of small plants, called *epiphytes,* growing on the branches. Some are beautiful orchids. They absorb water from the moist air, or trap rain in bucket-shaped leaves.

Temperate Rain Forests

Several parts of the world experience very high rainfall without the extremely high temperatures found in the tropics. These areas will support temperate rain forest, as on the west coast of North America, the west coast of New Zealand, and along the coast of southern Chile. The trees are nearly all cone-bearing species and include the largest of all trees, the giant redwoods and sequoias, which reach well over 100 metres in height. Epiphytes are abundant in these forests, but they are mainly ferns and mosses. Evaporation is not rapid because of the lower temperatures, so the forest literally drips with water. The low temperatures also slow down the rate of decay.

The jungle fowl is a native of South-East Asia. It is the ancestor of domestic chickens.

Animals of the Rain Forest

Monkeys, birds, fruit bats, amphibians, and snakes all feed in the canopy. Other animals feed on the forest floor, often on fallen fruit or soil-dwelling invertebrates. Rain forests are the home of monster insects, such as 30 centimetre centipedes, giant spiders and beetles, and wonderful butterflies.

Below: Tapirs are shy plant-eating forest dwellers of South America and South-East Asia. They live near to water and are excellent swimmers.

Mountain Life

As height increases, the temperature and the air pressure fall. These changes in physical conditions are reflected in the plant life. Luxuriant tropical rain forest at sea-level gives way to forests of broad-leaved deciduous trees and then to the coniferous forests. Higher still there are only stunted shrubs and small plants. Finally, a region of permanent snow and ice is reached.

These different belts of plant life are just like those we find if we make a journey from the Equator to the polar regions. The similarity is due to the fall in temperature towards the poles and towards the summits of mountains.

Alpine Plant Life

Severe conditions allow only very specialized plants to grow on the upper slopes of the mountains. Lichens are common for they can grow almost anywhere. Grasses are common, too, and there are also many broad-leaved flowering plants which are collectively called alpines.

Alpine plants are small plants, generally forming mats or cushions very close to the ground. In this way they avoid the full blast of the wind. They have long roots, which are necessary for secure anchorage and to draw up water, which drains rapidly from the stony ground.

Mountain Animals

A few insects, such as the apollo butterflies, and spiders are found on the higher mountain slopes. The most conspicuous of the upland creatures are the large grazing animals, such as the yak, the chamois, and the various kinds of sheep and goats. During the summer they wander high on the mountains, but most of them move down into the forests for the winter. Their large size and their shaggy coats are a protection against the cold. These animals move about on the rocky slopes with remarkable ease.

Right: Rocky Mountain goats.

Below: The mountain avens is a typical alpine plant, forming mats or cushions of tough leaves on the rocky ground.

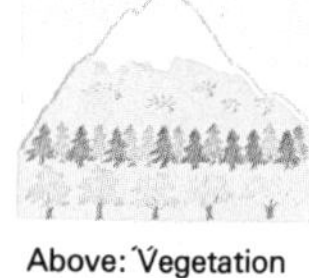

Above: Vegetation changes with altitude.

Desert Life

A desert is an area which receives less than 25 centimetres of rain per year. Although there are some deserts in the colder parts of the world, most deserts are in and around the tropics. Desert areas get extremely hot during the daytime, but they cool down rapidly at night. The plants and animals living in the deserts therefore have two main problems—shortage of water and great heat.

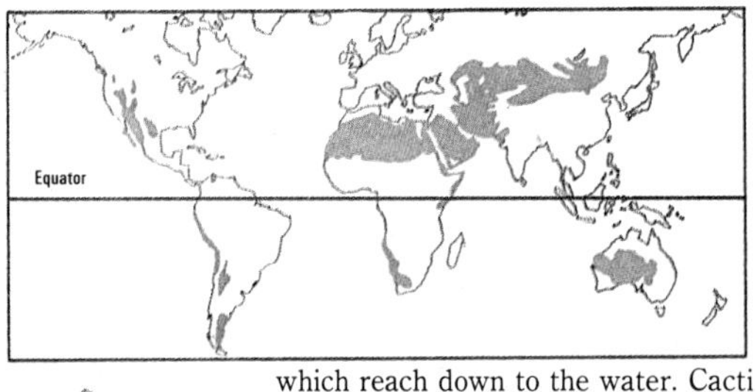

Desert Plant Life

Many of the smaller plants avoid the dry periods altogether as seeds. When it does rain, the seeds start to grow very rapidly. By the time the dry season sets in again they have already scattered a new crop of seeds.

The creosote bush of America can survive being shrivelled up, but most plants need a supply of water. The mesquite tree has very long roots which reach down to the water. Cacti store water in juicy stems.

Above: Nearly all desert regions have some form of jumping rodent. They have long tails which they use as rudders when leaping. This is a jerboa.

Desert Animals

The rains bring out the insects. Bugs, beetles, butterflies, and others emerge from their eggs or pupae. Within a short time the cycle is complete and another generation of eggs and pupae is left. Many insects will be eaten by lizards, snakes, scorpions, and mammals.

Frogs and toads have even adapted to the desert. The eggs hatch almost at once in temporary puddles and emerge as little toads in three or four weeks.

Few birds live in the desert. Seed-eaters need water and those that do live there seek the shelter of shrubs and rocks. Woodpeckers and owls use hollows in cacti.

Among the desert mammals, the most abundant are the rodents. Many never have to drink. They get all their water from their food, by conserving it very carefully. They sweat very little and pass very little urine.

The cacti are the best known of the desert plants. The great saguaros tower up to 20 metres above the desert floor. The spiny cacti provide a safe refuge for many small animals such as the cactus wren (below right).

Life in Northern Forests

Cool forests occur mainly in the Northern Hemisphere. They are of two main types: deciduous and coniferous. Coniferous forests grow mainly in the colder parts. The two types differ a great deal in their animal and plant life.

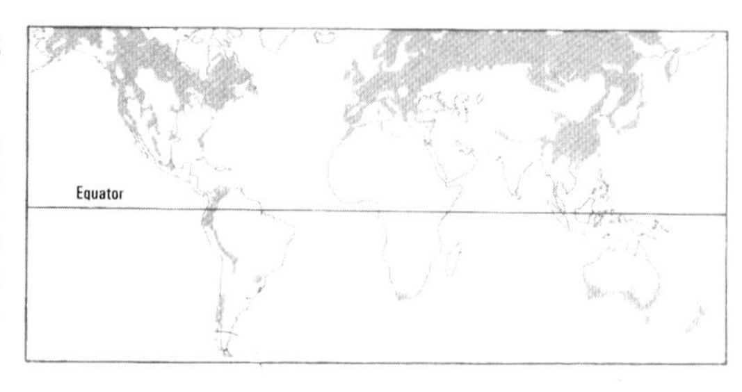

Distribution of coniferous and broad-leaved forests.

Deciduous Forests

Deciduous forests are composed mainly of trees that drop their leaves in the autumn. There is plenty of light in the spring, before the trees have spread their leaves so a rich variety of plants grows beneath them. These depend on the nature of the soil, but there are usually masses of flowers.

Dead leaves decay rapidly to form a layer of leaf litter. This is full of small invertebrate animals. Shrews, hedgehogs, mice, and voles are very common, feeding on the animals and plants.

In the trees, hordes of caterpillars and other insects are feeding on the leaves. Beetles and other wood-borers eat into the wood. These are snapped up by many birds. Other birds eat seeds and buds. Then there are the predatory hawks and owls. Foxes, badgers, boar, and deer wander through the trees.

Coniferous Forests

Coniferous forests grow in the colder parts of the world. Most of the trees are evergreen. Decay is slow on the forest floor and rain water removes useful minerals so the soil becomes poor. Little light penetrates and few plants grow under the trees, so the forests are poor in animal life. Insects on the leaves and in the bark attract plenty of birds, and others feed on seeds in the cones. The crossbill's strangely crossed beak is adapted for opening cones. Owls and other birds of prey are also plentiful in the forests.

Mammals are less common because there is less to eat in the leaf litter. There are shrews and voles, and many squirrels including some which live on the ground. Also in the trees there are martens and wild cats. Deer are common, and there are bears, wolves, and wolverines where they have not been hunted into extinction.

Squirrels are ideally adapted to a forest life. Sharp, curved claws enable them to climb trees rapidly and powerful back legs help them to leap from tree to tree using their outstretched tail to steer.

Woodpeckers (left) feed on insect larvae. They drill holes in the bark of trees with their chisel-like beaks and scoop up the larvae with their sticky tongues. Nuthatches (right) wedge hazel nuts in bark crevices and hammer them open with their beaks.

Life on the Seashore

Conditions on seashores are constantly changing. For part of the time they are covered with salt water. At other times they are exposed to the air and fresh rain water. They also are battered by waves. Despite these problems, many plants and animals make their home on the seashore.

The type of shore depends very much on the rocks and geography. Rocky shores support the most varied life because seaweeds can grow on them.

Above: The common puffin, a member of the auk family, is 20 centimetres long. It can hold several fish at once in its broad beak.

The Tides

The true seashore lies between the highest and lowest tide lines. Tides are caused by the pull of the Moon and, to a lesser extent, the Sun. When Sun and Moon pull together, *spring tides* result. These are higher and lower than normal. *Neap tides* are weak tides a fortnight later.

The Zones of the Shore

Many animals visit the shore: fishes when the tide is in and birds when it is out. Those that live there permanently are greatly affected by the tides. They are basically sea creatures whose main problem is to survive when the tide goes out. Some can withstand exposure better than others, so they separate into zones down the shore. This is best seen on rocky shores.

Above: The king shag, a New Zealand member of the cormorant family, is 50 centimetres long.

Right: The bar-tailed godwit (top) and the spotted sandpiper (bottom) are wading birds. In summer the male godwit is chestnut in colour, but in winter it becomes much paler.

The Splash Zone

This is above the high water mark, but it is splashed by spray. There are no seaweeds but lichens and flowers grow there, together with winkles and the sea-slater.

The Upper Shore

This is the region between the high tides of neap and spring tides. It is covered with slippery green seaweed where a few animals can live.

The brown pelican (top) lives in North America. It is 125 centimetres long. The common cormorant (below) is 90 centimetres long.

The Middle Shore

Between high and low tide marks of the neaps, the shore is covered and uncovered every day. It is rich in brown seaweeds that hide winkles, limpets, and many other animals.

The Lower Shore

This is covered most of the time, so sea anemones, mussels, and worms can live here.

Sandy Shores

These look empty but huge numbers of worms and shellfish live under the sand.

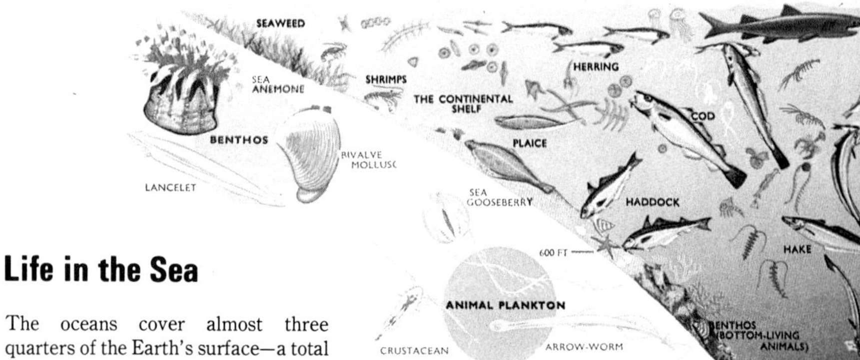

Life in the Sea

The oceans cover almost three quarters of the Earth's surface—a total of about 58 million square kilometres. The average depth is more than 3 kilometres and this means that the oceans contain something like 1200 million cubic kilometres of water. This immense volume of water supports a fantastic array of animal life, from the surface right down to the deepest trenches. All the major animal groups are represented in the oceans, but plant life in the ocean consists almost entirely of algae and bacteria.

The Pastures of the Sea

Except in the coastal regions, where seaweeds can grow on the bottom, all life in the ocean depends upon the tiny plants that float in the surface layers. These plants, together with the little animals that float with them, make up the *plankton.* Most of the planktonic plants are confined to the upper layers of the ocean.

The planktonic plants—collectively called *phytoplankton*—carry out photosynthesis just as land plants do, and they provide food for all the animals in the sea. Because of this, the surface layers are often called the pastures of the sea. These pastures are incredibly rich, and a given area can produce far more plant life than a similar area on land.

The phytoplankton consists entirely of minute plants called algae. There are several different kinds, but all of them are under one tenth of a millimetre across. The most common kinds are called *diatoms.* Each diatom is a single-celled plant and it lives inside a glassy box which is often beautifully sculptured. Diatoms are generally brownish green or yellowish.

Below: The Portuguese man o' war, a distant relative of the jellyfishes, has numerous long, stinging tentacles hanging down from a gas-filled float. Its poison is very powerful.

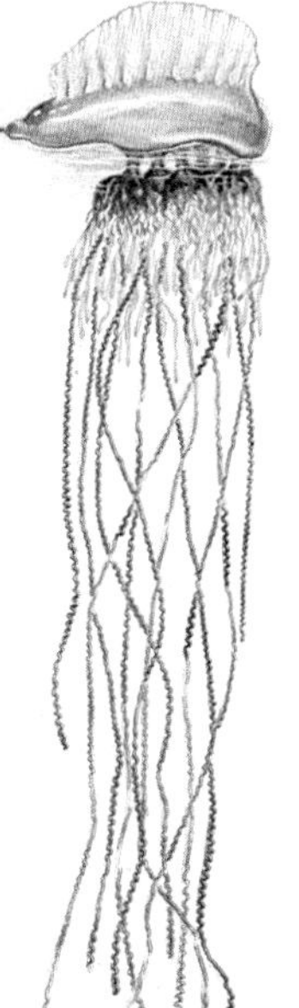

The animals of the plankton include a great many different types. Most of them are crustaceans, but there are also arrow worms, jellyfishes, and sea gooseberries. Many of these live permanently in the plankton, but others spend only part of their lives drifting at the surface. These 'part-time drifters' include the young stages of many crabs, barnacles, snails, bivalves, and starfishes. Planktonic animals often extend into deeper waters than the plants and many of them make daily journeys from one layer to another (see page 53). In common with the phytoplankton, many of the animals possess strange spines or feathery outgrowths. These

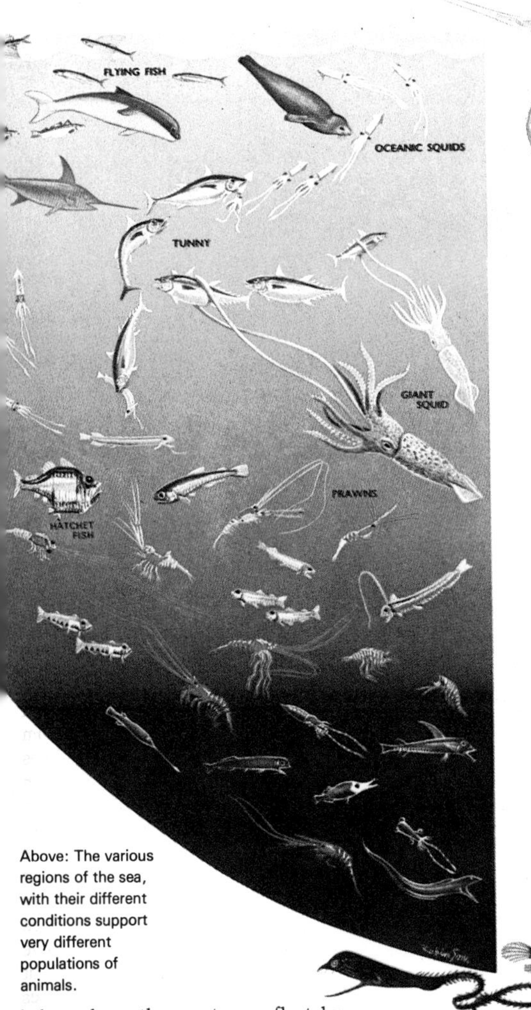

Above: The various regions of the sea, with their different conditions support very different populations of animals.

help to keep the creatures afloat by increasing the resistance to the water.

The planktonic animals feed upon the phytoplankton and upon each other. They in turn are eaten by larger creatures, such as herrings and other fishes.

The Swimmers

The active swimmers in the sea are collectively called the *nekton* and they

PLANKTONIC PLANTS

include fishes, squids, whales, prawns, and a few other creatures. Some of them live at the surface and feed directly on the plankton, while others live lower down. The whalebone whales feed directly on the small planktonic animals. They filter them from the water with great comb-like plates of baleen or 'whalebone' which hang in their mouths.

The Ocean Depths

The deeper the water, the darker it becomes. Below about 600 metres it is completely black and these deep waters contain a number of very strange fishes, squids, and prawns. Many of them have light-producing organs (see page 91). Some of the smaller animals feed on the debris falling from above, but most of the deep-sea creatures are carnivorous. Many of the fishes have huge mouths and formidable teeth.

The Ocean Floor

The worms, starfishes, sea anemones, and other creatures that live on

SOME DEEP-SEA FISHES

the sea-bed are collectively called the *benthos*. The bottom-dwelling creatures extend from the shore right down to the ocean depths, but their numbers get fewer as they get deeper. This is because food gets scarce. Most of the bottom-dwelling animals feed on the decaying matter that falls like rain from above.

Life in Fresh Water

Fresh water has been colonized from two directions: from the land and upriver from the sea. The major problem of living in water is to obtain oxygen. Animals coming from the sea had already solved this problem. They had the difficulty of adapting to the very different chemical conditions of fresh water.

Above: The kingfisher is a very striking bird of pond and stream. It sits on a branch overhanging the water and dives after fishes.

Below: The marshy ground around ponds and streams supports a number of characteristic plants.

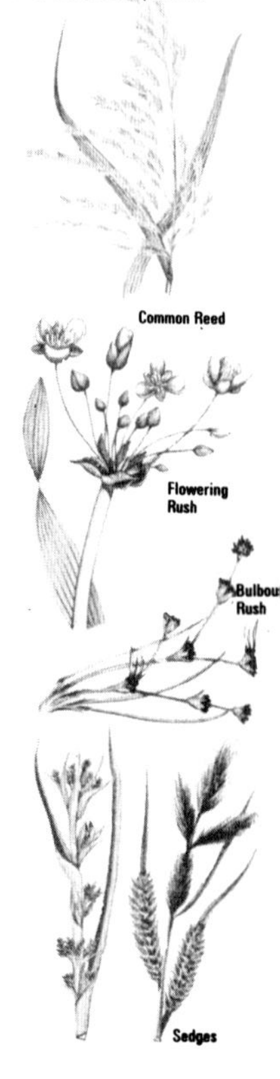

Water Plants

Fresh waters with sloping banks will be surrounded by lush vegetation. There will be reeds, rushes, and sedges with their roots in wet soil, but the real aquatic plants live in the water. They may be rooted, with floating leaves (e.g. water lilies) or with submerged leaves (e.g. Canadian pondweed). Floating plankton is also very important.

Water Animals

Nearly every group of animals can be found in fresh water, and every part of a pond has its characteristic inhabitants. The surface acts as a thin skin and supports pond-skaters and whirligig beetles.

Among the free-swimming creatures, there are fishes, water insects, water spiders, and mites, and a horde of little crustaceans.

Living in or on the mud, or crawling on the water plants, are the snails and mussels, insect larvae, and worms.

Life in Streams

The moving waters of streams and rivers present some different living conditions from those in still waters of lakes and ponds. It is possible to divide the stream into regions according to the plant and animal life.

The upper reaches form the *headstream*, where water is fast and shallow. There is plenty of oxygen, and the main problem is to avoid being swept away. The animals and plants are designed to cling tightly.

The headstream merges with the *troutbeck*. The slope is less steep, but it is still rocky and fast-flowing. Fishes appear, either as strong swimmers, like trout, or hiding under stones, like loach.

The next stretch is the *minnow reach*. Sand and gravel covers the bed and there are now plenty of water plants. Finally, there is the deep, slow *cyprinoid reach*, with a muddy bottom, plenty of plants and many fishes of the carp family, such as rudd and roach.

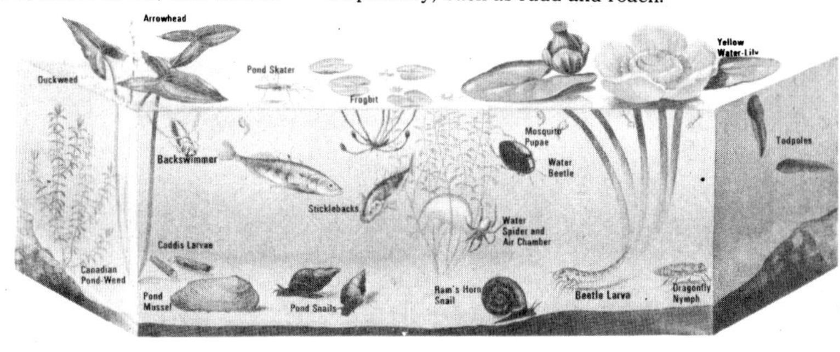

Polar Life

Because the Earth is tilted on its axis, the polar regions experience at least one day in summer when the sun never sets, and one day in winter when it never rises. Polar winters are intensely cold, and even in the summer the temperature may stay below freezing because the sun's rays strike at a low angle, and much of the heat is reflected by the snow and ice.

The main difference between the Arctic and Antarctic regions is that the Arctic is an ocean surrounded by land, whereas the Antarctic is a continent, larger than Europe, surrounded by ocean.

Life in the Arctic

Between the *tree line,* beyond which no trees grow, and the shores of the Arctic Ocean lies the *tundra*—a vast expanse of flat land which is frozen and bare for most of the year.

The most abundant Arctic plants are lichens, which encrust bare rock surfaces and may cover large areas of ground. Reindeer moss, which is eaten by reindeer, is a lichen. Dwarf birch and willow survive for hundreds of years, but they grow very slowly. In the summer there are numerous flowering herbs.

Animals are rarely seen on the tundra in winter, but they reappear when the snow melts. Insects are soon on the wing, and lemmings appear from their winter burrows. Ground squirrels awake from hibernation, while Caribou and wolves move north from the forests. Ducks, waders, and other birds fly in from the south to nest on the boggy tundra.

A nesting skua shrieks a warning in Antarctica. Skuas are rapacious birds. In the Antarctic their diet consists mainly of the eggs and young of penguins.

The snowy owl hunts over the Arctic tundra. Small animals, such as lemmings, form its chief food.

Life in the Antarctic

Most of the Antarctic continent is permanently covered with ice. Only the coast thaws out during the summer. The only plant life is some 150 species of lichens, a few mosses, but only three species of flowering plants. This sparse vegetation supports only very little animal life, and the only truly terrestrial animal is a wingless fly about 5 millimetres long.

The more characteristic animals live in the sea. They include seals, penguins, other seabirds, and whales. Except for the whales, they come out of the water to breed, but they get all their food from the sea in the form of fish, squid, and crustaceans. The emperor penguin breeds on the frozen sea. The male carries the egg on its feet for eight or nine weeks in the winter until relieved by the female.

Life on the Grasslands

The major grasslands of the world occur on each side of the desert belts, in regions where there is a certain amount of rain but not enough for the growth of forests. There are two main types of grassland: temperate and tropical.

Temperate Grasslands

The temperate grasslands are found in the inner regions of the continents, where rainfall is low throughout the year. Among the major temperate grassland areas are the prairies of North America, the pampas of South America, the veld of South Africa, and the steppes of Eurasia. Smaller areas are found in eastern Australia.

Where the rainfall is moderately high, the grasses are tall and lush. A large proportion of the temperate grassland has been taken over for agriculture and stock rearing.

Tropical Grasslands

Tropical grasslands, also called savanna, are between the tropical forests and the deserts. The largest areas are in Africa and South America, and smaller areas occur in India and northern Australia. Rain falls only for a short period during the summer. The rain and the high temperature produce rapid growth and the land quickly becomes covered with lush grass. Then, as the dry season progresss, the plants die down and the land takes on a rather barren appearance.

Above: The sable antelope of African grasslands.
Below: The world's grasslands.

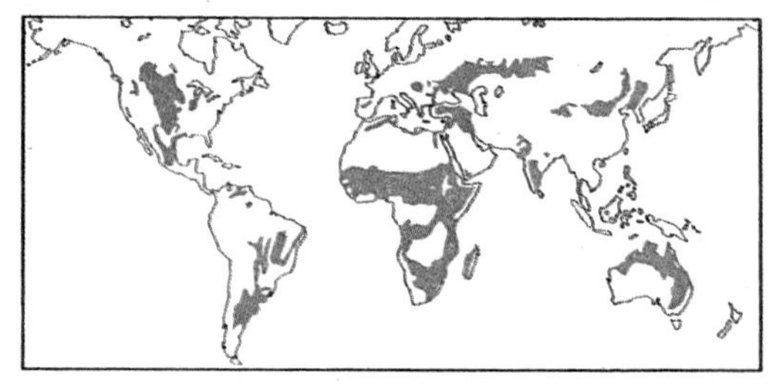

The Animals of the Grasslands

The dominant animals of the grasslands are the large grazers. These include horses, antelopes, zebras, bison, and kangaroos. The savanna lands, where there are trees and bushes, also support elephants, giraffes, and rhinos. Grasslands do not provide much cover for the animals, and we find that most of the large grazers are fast runners. This is their way of escaping from their enemies. Most of them also live in herds.

The grasslands also support many smaller mammals, such as voles and prairie dogs. There are also many insects, notably the grasshoppers and locusts. These feed on the grass and other plants and they themselves are eaten by birds, lizards, and small mammals.

Best known of the carnivorous mammals of the grasslands are the various large cats—the lions and leopards, for instance. The lions and leopards of Africa feed mainly on the antelopes and zebras. Another famous

The elephant is found over much of Africa, in both grassland and forest. It is entirely vegetarian.

Above: The water hole is very important for the grazing animals. Zebras and wildebeest have come to drink but some individuals are always on the look-out for danger.

Right: A pride of lions. Lions are rather lazy animals and spend a lot of time lying in the shade. They kill only when they are hungry.

Right: The coyote or prairie wolf of the American prairies. It feeds mainly on hares and small rodents.

Left: Termite nests are a dominant feature of some areas of tropical grassland. Made of soil cemented by saliva, they may rise 6 metres above the ground.

hunter is the cheetah. The fastest land animal in the world, the cheetah can reach 100 kph as it bounds after its prey. It cannot keep this up for long, however, and it gives up if it does not catch its prey within about 400 metres. Other well known hunters include the coyote or prairie wolf of North America and the hunting dog of southern Africa.

Water is always a problem to animals living in the grasslands, especially those in the tropical grasslands where rain falls for only a short time each year. Water holes are very important. Many animals come to drink daily and they often come at definite times, each species or group of species coming in turn. When the water holes eventually dry up, as they often do towards the end of the dry season, the herds move on to other regions. One hundred years ago there used to be mass migrations of antelopes in herds many kilometres long. There are far fewer animals now, although one can still see migrating processions. The smaller animals often sleep in burrows for the driest part of the year. In the temperate grasslands, where the winters are very cold, many of the smaller animals hibernate.

Fossils

Fossils are the traces and remains of animals which have been naturally preserved in the rocks, sometimes for many millions of years.

The rocks are like the pages of a history book to a geologist: they tell the story of the Earth. But they are far more difficult to read than the pages of an ordinary book because they are frequently torn, bent, or upside-down. Some of the pages are scattered over a wide area, and some are missing altogether. The key to the pages and their order lies in the fossils contained within the rocks.

If we take an undisturbed sequence of rocks it is fairly obvious that the oldest rocks will be at the bottom and the newest rocks, which were laid down most recently, will be at the top. If we look at the fossils in such a sequence of rocks we will find that the fossils at one level are not exactly the same as those at another level. As one goes higher up the sequence, some of the fossils disappear, and new kinds come in to take their place. Each layer of rock thus has a characteristic collection of fossils.

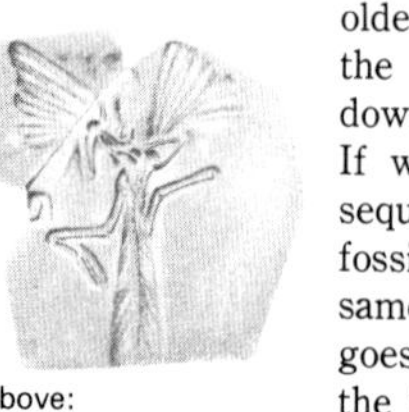

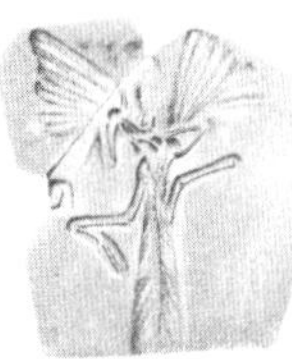

Above: *Archaeopteryx*, the first known bird, is known from fossils found in a Bavarian limestone quarry.

Right: Fossil crinoid or sea lily.

Below: Fossil of *Paradoxides*, the 50-centimetre giant trilobite of the Cambrian seas.

This is of great importance to the geologist because if he finds the same assemblage of fossils in rocks from different regions he can safely say that the rocks are of about the same age.

But as well as showing us some of the Earth's history, the fossils in the rocks show us how the plant and animal life of our planet has changed. 500 million years ago there were plenty of shelled creatures related to today's crabs and cuttlefishes. There were also many other kinds of animals that are now completely extinct, but there were no animals with backbones as far as we know.

Then, about 450 million years ago, perhaps, the first fishes appeared. Only fragments of these early fishes have been found but we can be fairly certain that the backboned creatures had arrived on the scene by this time. 100 million years later, the fossil record tells us that a great number of different kinds of fishes were in existence. Soon after this amphibians began to live on the Earth. Still later, about 300 million years ago, the reptiles appeared. Birds and mammals did not appear until much later.

The fossils in the rocks therefore show us roughly when the various groups of animals and plants came into being. But they do more than this. There are some rare fossils that do not fit easily into any of the main groups. Some of them look partly like fishes and partly like amphibians; others look partly like amphibians and partly like reptiles. These fossils show us that the fishes—or, to be more accurate, some fishes—gradually changed into amphibians, and that some amphibians gradually changed into reptiles. In other words, the fossils show us that life *evolved* and that all today's plants and animals have developed from simpler forms of life.

Looking for Fossils

Fossils are quite easy to find if you know the sort of places in which to look for them. You can often pick them up in fields and gardens, but these scattered finds are often incomplete and damaged because they have been rolling around for a long time. Better specimens are obtained from freshly exposed rocks. Granite and similar rocks contain no fossils because these rocks formed from molten material.

If you want to find fossils you must look at the sedimentary rocks, which formed from sediments deposited in the seas and lakes. Sandy rocks are not very good for fossils and rarely contain complete specimens. Limestones, clays, and shales are the best rocks to examine. Many limestones consist almost entirely of fossils. The fossils are often harder than the surrounding rock and they stand out from the surface. They can then be chipped out with a hammer and cold chisel. Always examine the heaps of clay and other rocks brought up during roadworks. Railway cuttings, quarries, and sea cliffs are also very good places to look for fossils.

THE AGES OF THE EARTH

Geological Periods	Plant and Animal Life Forms
Quaternary Period Began 2-3 Million Years Ago	Great Ice Age. Coming of Man.
Tertiary Period Began 70 Million Years Ago	Modern Mammals Appear.
Cretaceous Period Began 135 Million Years Ago	Dinosaurs Decline. Mammals Increase. First Flowering Plants.
Jurassic Period Began 180 Million Years Ago	Dinosaurs Abundant. Birds Appear.
Triassic Period Began 225 Million Years Ago	Early Dinosaurs. The First Mammals.
Permian Period Began 270 Million Years Ago	Reptiles Increase.
Carboniferous Period Began 350 Million Years Ago	First Reptiles.
Devonian Period Began 400 Million Years Ago	Age of Fishes. First Amphibians.
Silurian Period Began 440 Million Years Ago	Earliest Land Plants.
Ordovician Period Began 500 Million Years Ago	First Fishes.
Cambrian Period Began 600 Million Years Ago	First Abundant Fossils, All Invertebrates.

Formation of Earth About 4,500 Million Years Ago

Mammoths have been found preserved in the frozen wastes of Siberia and Alaska.

How a Fossil Forms

When an animal dies its body decomposes or is eaten by other animals. But the hard parts—the teeth and the bones or shell—are not so easy to destroy, and if they are buried quickly they stand a good chance of being fossilized. Most fossils that have been discovered are of marine creatures, because the best conditions for fossilization occur in the sea.

When the animals died their shells or skeletons became buried under a rain of sand and mud. In time the sediment became compressed and hardened into rock and the animal remains became converted into fossils. At a much later date, movements of the Earth's crust lifted up these new rocks to form land. Wind and rain began to wear down the rocks and the fossils eventually came to light, in a riverbank or seacliff.

Very occasionally the actual skeleton may be preserved. This has happened where animals have been trapped in bogs or tar pits and buried very rapidly. The Californian tar pits, for instance, have yielded a wealth of animal skeletons including those of several birds.

The buried remains undergo a certain amount of change. The skeletons and shells are most commonly *petrified,* or converted into stone. Water percolating through the rocks gradually dissolves the original material and deposits mineral matter in its place. If the original material is dissolved away and not replaced it leaves a hollow in the rock. This hollow

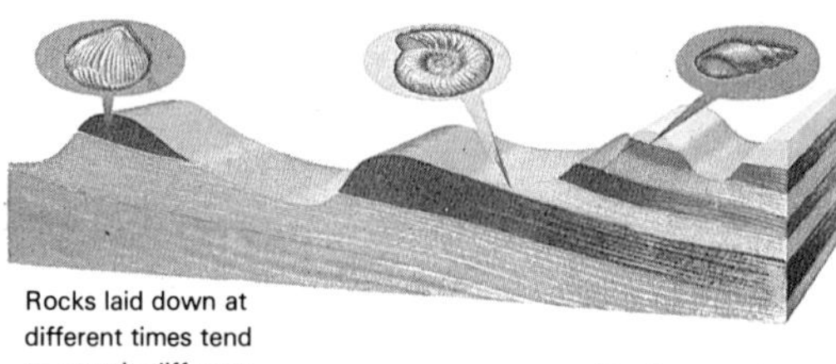

Rocks laid down at different times tend to contain different collections of fossils which enable the rocks to be recognized wherever they are found.

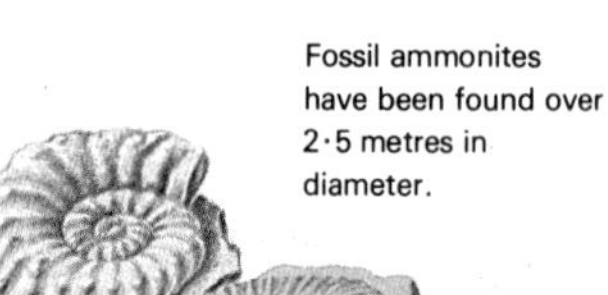

Fossil ammonites have been found over 2·5 metres in diameter.

Fossil remains of insects have been found in amber, the hardened resin of evergreen trees.

FOSSILS

Ammonites

The ammonites are among the best-known fossils. They were marine creatures related to today's squids and cuttlefishes. The original shell has almost always disappeared, and the fossils are either moulds or internal fillings. The internal fillings were formed when the shells became filled with sediment and then dissolved away, leaving a detailed pattern of the inside of the shell. Some of the fossils in the clays are composed of iron pyrites and they sometimes shine like gold.

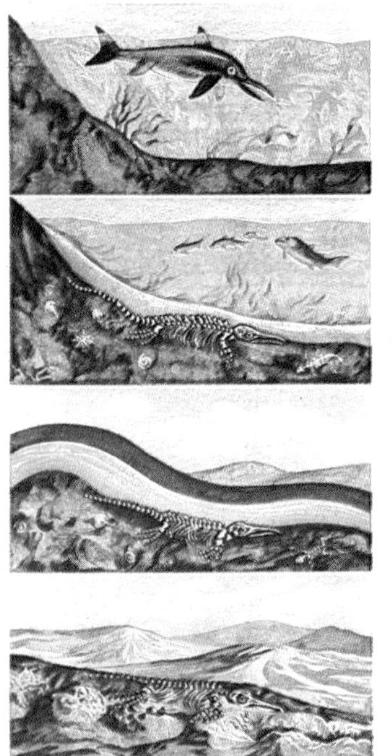

Diagram showing how an animal may be fossilized. A sea-dwelling ichthyosaur dies and sinks to the sea-bed. Soon its skeleton is covered with mud which gradually settles into solid rock. At a much later date earth movements buckle the sea-bed and cause it to rise above sea-level. Erosion then strips away the rock covering until the fossil is exposed.

is called a *mould* and it is a very common form of fossil. If mineral matter is deposited in the mould it forms a natural *cast*. This type of fossil shows all the external features of the object, but it does not tell us anything about the internal structure of the object.

Many plant fossils are simply residues of carbon, which give the actual shapes of the leaves and stems. Some animals may be preserved in this way, too. Pieces of amber, which is a hardened or fossilized resin from coniferous trees, sometimes contain the remains of insects which had become trapped in the sticky resin.

Ancient animals sometimes left tracks and footprints in soft mud which became hard. The prints then became filled with some other material and preserved as fossils. They tell us quite a lot about the animal that made them. Although a footprint preserved in rock is not actually part of an animal, it is a true fossil and it does tell us something about the life of the animal that made it.

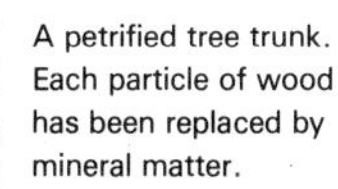

A petrified tree trunk. Each particle of wood has been replaced by mineral matter.

Unusual fossils are tracks left by prehistoric animals in mud which later hardened into solid rock.

Sometimes the fossilized skeleton of an animal may be surrounded by a film of carbon showing the actual outline of the animal's body.

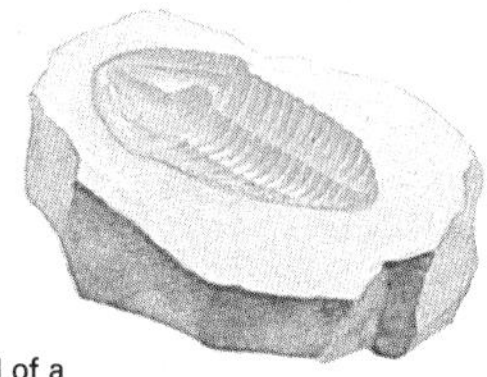

A mould of a trilobite, the impression that may remain after the animal itself has disappeared.

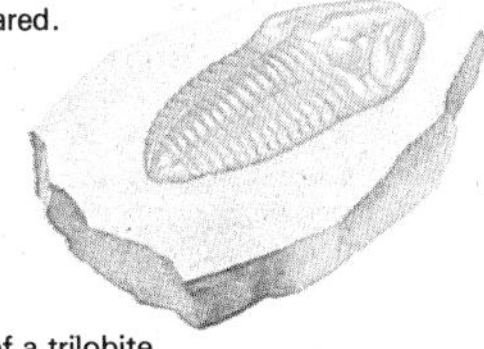

A cast of a trilobite formed by the later deposition of mineral matter in the original mould.

Below: A fossil dragonfly.

The Theory of Evolution

Evolution means the gradual change of one type of plant or animal into another. It is generally believed that a primitive form of life began on the Earth hundreds of millions of years ago and that all of today's plants and animals have evolved from this simple ancestor. The evidence to support this idea of the evolution of living things comes from several directions.

Evidence from Fossils

Fossils provide some of the strongest support for the idea of evolution, as we have already seen in the last few pages. *Archaeopteryx* was a primitive bird, but it had many reptilian features. This strongly suggests that birds evolved from reptiles. There is even stronger evidence concerning the development of the horse. There are fossils showing almost every stage between a little fox-sized animal and a modern horse. The explanation is that the animals have gradually changed during the last 60 million years. The changes from one generation to the next would have been imperceptible, but over millions of years they added up.

Evidence from Anatomy

A whale's flipper, a bird's wing and a man's arm look different, and they are

Convergent Evolution

The process of natural selection ensures that a plant or an animal develops the most efficient shape and structure for its particular way of life. Animals leading similar lives may thus come to resemble each other very closely, although they may be quite unrelated. This is known as convergent evolution. A good example is the European mole and the marsupial mole of Australia. They are remarkably similar, yet one is a pouched mammal and the other is a placental belonging to a very different group of mammals.

Genetics and Evolution

Every cell in the body contains a number of minute thread-like structures called *chromosomes*. Each one carries a number of *genes*. The genes are the instructions which ensure that the cells (and therefore the whole body) develop in the right way. When plants or animals reproduce they normally combine instructions from both parents. Slightly different combinations of genes are produced in the offspring, and these give rise to slight variations. Natural selection then favours those which are useful. The species thus becomes more efficient and better adapted to its surroundings. Sometimes there is a 'mistake', or mutation, when the instructions are handed on to the offspring. Wrong instructions are usually harmful and the organism usually dies, but sometimes they result in major improvements. Organisms with a useful mistake will survive and breed. The useful feature will thus be handed on.

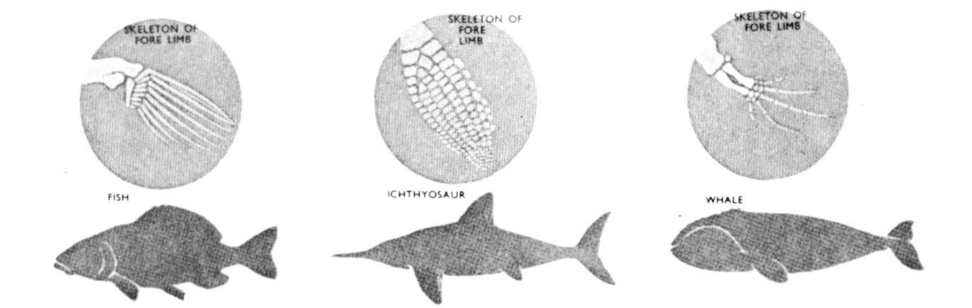

A fish, an ichthyosaur (reptile), and a whale (mammal) – three types of swimming vertebrate. The fore-limbs are homologous (similar in relative position and structure) showing that the three groups have an ancestry, however remote, in common.

used for different jobs, but the bones are remarkably similar. This suggests very strongly that the animals have evolved from the same stock and that they have merely adapted the original skeleton for different jobs.

Evidence from Embryos

Lizards, birds, rabbits, and men all look very different. In their early stages, however, they all look rather like fishes. The explanation is that they have all descended from some kind of ancestral fish.

Young animals also provide evidence for evolution among the invertebrates. Some marine worms and molluscs have very similar young stages. This suggests that the two groups, very different in adult form, have evolved from a single ancestral group.

An Old Idea

The idea that living things have evolved is not new. The Greeks suggested it about 2,500 years ago, but they had little evidence to back up the suggestion at the time. The idea cropped up again from time to time during the last thousand years, but most people preferred to believe in special creation. They believed that each kind of plant and animal was created by some supernatural power and that it did not change throughout its existence.

One of the main reasons why the idea of evolution was not readily accepted was that no-one could explain how or why the evolutionary changes might have taken place. Charles Darwin provided the answers in the middle of the 19th Century and the idea of evolution gradually came to be accepted.

Darwin's Voyage on the Beagle

In 1831, when he was only 22 years old, Charles Darwin began a round-the-world voyage in the survey ship *H.M.S. Beagle*. When the ship left England Darwin believed in special creation, but as the voyage went on he began to have doubts. He saw that animals and plants were beautifully adapted to their surroundings, but he also noticed that the animals and plants

Darwin discovered a dozen or so species of finch inhabiting the Galapagos Islands of the Pacific. The ancestral species of finch is believed to have come from South America, 966 km to the east. In different parts of the archipelago new distinct species evolved from this original type, each adapted to a particular form of feeding.

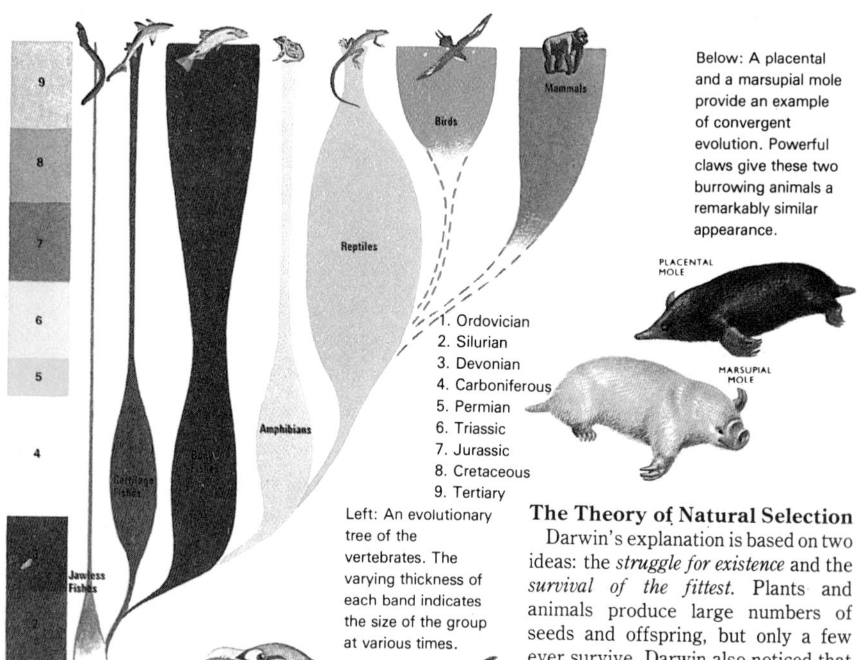

Below: A placental and a marsupial mole provide an example of convergent evolution. Powerful claws give these two burrowing animals a remarkably similar appearance.

Left: An evolutionary tree of the vertebrates. The varying thickness of each band indicates the size of the group at various times.

The flightless great auk became extinct in the last century.

in one part of the world were not the same as those in similar habitats elsewhere. Why should so many types have been created to fill one type of habitat? In the Galapagos Islands he found numerous finches. They all had a general resemblance to each other and to the finches of the South American mainland, yet they were clearly different species. Why should so many species have been created in this small area, and why should they have so many similarities? Darwin came to the conclusion that the various species had not been created individually, but that they had all descended from a common ancestor. In other words, he came to the conclusion that evolution had occurred. He then looked for an explanation.

The Theory of Natural Selection

Darwin's explanation is based on two ideas: the *struggle for existence* and the *survival of the fittest.* Plants and animals produce large numbers of seeds and offspring, but only a few ever survive. Darwin also noticed that the members of a species vary among themselves. He reasoned that the individuals with the most useful variations would survive best. These would be the ones that would breed, so the next generation would possess the useful variations. There would still be a struggle for existence, and again the fittest would survive. This process would go on generation after generation and the plants or animals would gradually get better suited to their surroundings. They would get more and more efficient at catching

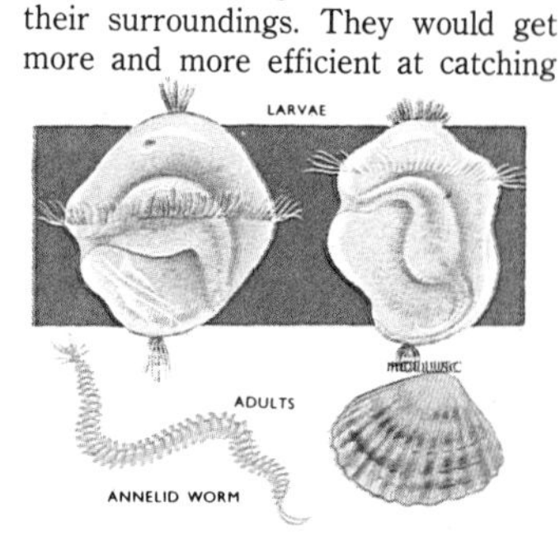

Right: Evidence that molluscs and annelid worms had a common ancestor is provided by the existence of similar larval forms in some species.

food and escaping from their enemies. Conditions are not the same everywhere, and so a variation that was useful in one place might not be useful in another. A given species might therefore change in two or more directions and give rise to two or more different species, each well suited to its own habitat. Darwin called his idea the Theory of Natural Selection, and we now believe that all plants and animals have evolved and become adapted to their surroundings by this method.

Evolution Today

It must not be thought that evolution is something that took place in the past and then stopped. It is taking place today just as much as ever, and it will continue for as long as conditions continue to change. One of the best examples of evolution that we can see taking place today concerns the spread of black (melanic) varieties of certain moths. The peppered moth is one of the best known. The normal form of this insect is black and white, but completely black individuals used to crop up every now and then—just as white (albino) varieties occur in other animals.

Until about 1850 the black moths were rare and they did not get much chance to breed. The moths rest on tree-trunks by day, and the normal forms were protected by camouflage among the lichens. The black ones, however, were easily spotted by birds and eaten. When the industrial revolution came and smoke began to blacken the tree-trunks, the black moths had more of a chance. They began to breed and get commoner. The black ones were now better camouflaged on the trees than the normal speckled forms. The original black-and-white form of the peppered moth is now quite rare in industrial regions. Man was responsible for the change in the surroundings, but the change in the moths came about entirely by natural selection.

The black (melanic) form of the peppered moth has increased during the last hundred years or so in industrial areas at the expense of the normal peppered moth because smoke pollution has blackened buildings and trees, thus providing it with better camouflage. The photograph below shows both moths on normal unblackened bark.

Living Fossils

In 1938 a strange fish was brought up by a trawler working off the coast of South Africa. The fish was heavily built and nearly 2 metres long. Its fins were most unusual because they were something of a cross between normal fish fins and the limbs of land animals. Scientists looked at it and realized that it was a coelacanth.

Scientists knew about coelacanths only as fossils. They were rather primitive fishes and they were common 350 million years ago, but they were thought to have become extinct more than 70 million years ago. It is not surprising that the coelacanths have been called 'living fossils'.

Why then should the coelacanth have survived? Although a primitive type of fish, it had become very well adapted to the conditions under which it lived, and the later fishes have failed to produce a more efficient species for those particular conditions. Conditions in the sea have not changed a great deal during the Earth's history, and the features evolved by the ancient animals are still useful today.

The tortoise belongs to a group of reptiles which have changed very little since they first appeared some 200 million years ago.

The coelacanth was rediscovered in 1938 after it was thought to have become extinct some 70 million years ago.

The tuatara is the only living representative of a group of reptiles more ancient than the dinosaurs.

There are many other marine animals that qualify for the title 'living fossils'. Examples include the nautilus and the king crab. The lung fishes are also living fossils, very similar to creatures that became extinct long ago.

It is not only in the sea that we find living fossils. The tortoises and crocodiles belong to very ancient groups of reptiles and they have changed very little since they first appeared some 200 million years ago.

Ancient animal types can also survive to become living fossils if they become isolated from the main centres of evolution. New Zealand, for example, is the home of the lizard-like tuatara. This reptile is about 60 centimetres long and it is only distantly related to the lizards. It is the only living representative of a group more ancient than the dinosaurs. New Zealand was separated from the rest of the world's land masses before the dinosaurs evolved, but some of these more ancient reptiles had already arrived there. Isolated from competition, they flourished here until the middle of the 19th century. The tuatara now lives only on a few off-shore islands, but it is quite plentiful there.

The Balance of Nature

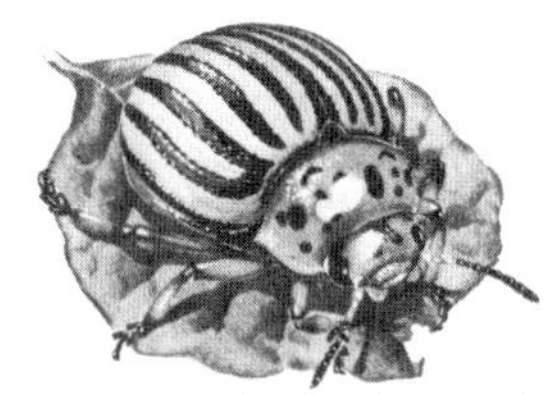

No animal lives alone: each one is intricately bound up with dozens of other organisms in the *web of life*.

Plants and animals live in communities and the members of each community have become perfectly adapted to each other. There are just the right number of plant-eating animals to keep the plant life in check without any risk of overgrazing. Similarly, there are just enough carnivores to balance their prey animals. The habitat as a whole remains unchanged over a long period of time.

Large-scale changes, however, would have a serious effect on the community. Removal of one species would destroy must of nature's web, because all the species of a community are linked up. Unfortunately, Man is doing just this in many places today.

Upsetting the Balance

When men learned to grow crops and domesticate animals, they settled in permanent homes. Clearing the land to make way for farming led to the loss of many wild animals and plants. Land clearance is still going on today, and at a faster rate to accommodate our increasing population.

Natural communities contain large numbers of different species. They are kept in check by one another and their numbers remain steady. Artificial communities do not have such stability. There are few species in a field of crops and one species may increase to enormous numbers. In other words, it becomes a pest.

The balance is also upset by accidental or deliberate introductions. These species have no natural enemies and multiply enormously at the expense of native species.

Artificial communities of plants do not support a large number of different species of animals. The result is that one species often increases to enormous numbers. In other words it becomes a pest. The Colorado beetle has ravaged potato crops in North America and Europe.

Removal of the natural vegetation has resulted in soil erosion in many parts of the world. The Tennessee Valley in North America was a particularly badly affected area. Now work is underway to repair the damage by replanting.